普通高等教育“十三五”规划教材

光电信息科学与工程类专业规划教材

光电信息科学与工程专业英语教程

（第2版）

张　彬　钟哲强　编著

電子工業出版社

Publishing House of Electronics Industry

北京 • BEIJING

内 容 简 介

本书主要介绍光电类专业基本词汇和语法，并结合有关几何光学、波动光学、信息光学及技术、激光原理及技术、光纤通信原理及技术、光电子器件、光电子技术等方面专业知识，培养学生专业文献阅读、翻译与科技论文写作的基本技能。

本书可作为光学、光信息科学与技术、光电子技术、光电信息工程等专业本科生和硕士生的专业英语教材，也可供从事相关专业工作的工程技术人员参考。

图书在版编目（CIP）数据

光电信息科学与工程专业英语教程 / 张彬，钟哲强编著. —2 版. —北京：电子工业出版社，2019.8
ISBN 978-7-121-36910-0

Ⅰ. ①光… Ⅱ. ①张… ②钟… Ⅲ. ①光电子技术－信息技术－英语－高等学校－教材 Ⅳ. ①TN2

中国版本图书馆 CIP 数据核字（2019）第 122763 号

责任编辑：韩同平
印　　刷：北京捷迅佳彩印刷有限公司
装　　订：北京捷迅佳彩印刷有限公司
出版发行：电子工业出版社
　　　　　北京市海淀区万寿路 173 信箱　　邮编：100036
开　　本：787×1092　1/16　印张：13.5　字数：449 千字
版　　次：2012 年 11 月第 1 版
　　　　　2019 年 8 月第 2 版
印　　次：2019 年 12 月第 2 次印刷
定　　价：49.90 元

凡所购买电子工业出版社图书有缺损问题，请向购买书店调换。若书店售缺，请与本社发行部联系，联系及邮购电话：（010）88254888，88258888。

质量投诉请发邮件至 zlts@phei.com.cn，盗版侵权举报请发邮件至 dbqq@phei.com.cn。

本书咨询联系方式：（010）88254525，hantp@phei.com.cn。

前　言

本书主要针对光学、光信息科学与技术、光电子技术、光电信息工程等专业本科生和硕士生开设专业英语课程而编写，旨在使学生掌握基础专业词汇，提高专业文献阅读和翻译能力，拓展和深化学生对本学科专业基础知识和相关技术的了解和认识。

全书由 7 个主题单元组成。

1. 几何光学知识和有关词汇。主要内容包括：反射和折射、成像、物像关系、透镜和成像分类、像差等。

2. 波动光学知识和有关词汇。主要内容包括：波、波的叠加、衍射、近场和远场、相干性、干涉、偏振等。

3. 信息光学及技术知识和有关词汇。主要内容包括：全息照相术原理、彩虹全息、数字全息、光信息处理、傅里叶变换光学、空间滤波等。

4. 激光原理及技术知识和有关词汇。主要内容包括：光放大、光学谐振腔、横模、纵模、高斯光束、脉冲放大、信号畸变、放大自发辐射、调 Q 技术、锁模技术、选模技术、频率控制技术、波长选择技术、激光应用技术等。

5. 光纤通信原理及技术知识和有关词汇。主要内容包括：光纤通信系统的发展、光纤结构、光纤损耗、光纤色散、光纤非线性、光纤中的光传输、光纤非线性的影响、全光网络等。

6. 光电子器件知识和有关词汇。主要内容包括：光学材料、光谱滤光片、集成光学元器件、光电探测器等。

7. 光电技术知识和词汇。主要内容包括：薄膜技术、光刻技术、光子生物学技术、显示技术、红外探测技术、增材制造、太赫兹技术及应用等。

每个单元后面列出了课文中新出现的专业词汇，以供学生积累专业词汇。此外，每个单元还介绍了科技英语阅读与翻译的相关知识，以帮助学生提高科技英语阅读和翻译能力。

书中课文主要选自相关专业的英文原版教材、专著及专业文献。

在本教材的编写过程中，作者的研究生黄人帅、凌芳、钟亚君同学付出了辛勤的汗水，在此表示衷心的感谢！

由于编者水平所限，书中难免有些疏漏和欠妥之处，敬请读者不吝赐教（zhangbinff@sohu.com）。

编　者

目　　录

Part 1　Ray Optics

In this part, we treat light beams as rays that propagate along straight lines, except at interfaces between dissimilar materials, where the rays may be bent or refracted. This approach, which had been assumed to be completely accurate before the discovery of the wave nature of light, leads to a great many useful results regarding lens optics and optical instruments.

1.1　Refraction and Reflection

1.1.1　Refraction

When a light ray strikes a smooth interface between two transparent media at an angle, it is refracted. Each medium may be characterized by an index of refraction *n*, which is a useful parameter for describing the sharpness of the refraction at the interface. The index of refraction of air (more precisely, of free space) is arbitrarily taken to be one, *n* is most conveniently regarded as a parameter whose value is determined by experiment. We know now that the physical significance of *n* is that the ratio of the velocity of light in vacuo to that in the medium.

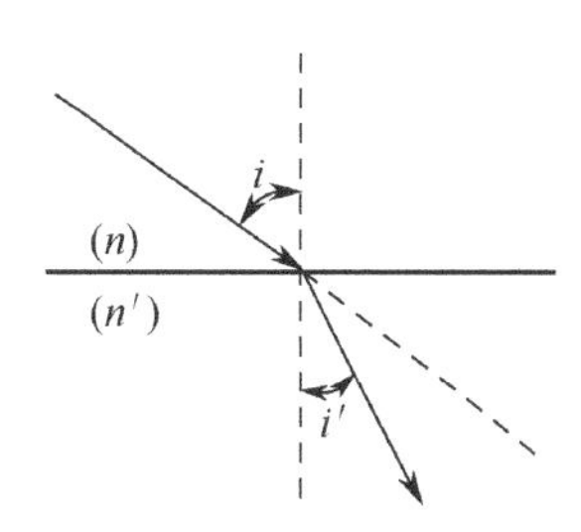

Fig.1.1　Refraction at an interface

Suppose that the ray is incident on the interface, as shown in Fig.1.1. It is refracted in such a way that

$$n\sin i = n'\sin i' \tag{1.1}$$

no matter what the inclination of the incident ray to the surface, *n* is the index of refraction of the first medium, n' that of the second. The angle of incidence *i* is the angle between the incident ray and the normal to the surface; the angle of refraction i' is the angle between the refracted ray and the normal.

1.1.2　Index of Refraction

Most common optical materials are transparent in the visible region of the spectrum, whose wavelength ranges from 400 to 700nm. They exhibit strong absorption at shorter wavelengths, usually 200nm and below.

The refractive index of a given material is not independent of wavelength, but generally increases slightly with decreasing wavelength (Near the absorption edge at 200 nm, the index of glass increases sharply). This phenomenon is known as dispersion. Dispersion can be used to display a spectrum with a prism; it also gives rise to unwanted variations of lens properties with wavelength. Table 1.1 gives typical index of refraction of several materials.

Tab.1.1 Index of refraction of several materials

Material	Index of refraction	Material	Index of refraction
air	1.0003	sodium chloride	1.54
water	1.33	light flint glass	1.57
magnesium fluoride	1.38	Sapphire	1.77
vitreous silica	1.46	extra-dense flint glass	1.73
Pyrex glass	1.47	carbon disulfide	1.62
Methanol	1.33	zinc sulfide (thin film)	2.3
xylene	1.50	medium flint glass	1.63
ethanol	1.36	titanium dioxide (thin film)	2.4~2.9
crown glass	1.52	heaviest flint glass	1.89
benzene	1.50	Canada balsam (center)	1.53

Optical glasses are generally specified both by index n and by a quantity known as dispersion v,

$$v = \frac{n_F - n_C}{n_D - 1} \tag{1.2}$$

The subscripts F, D and C refer to the indexes at certain short, middle and long wavelengths (blue, yellow, red).

1.1.3 Reflection

Certain highly polished metal surfaces and other interfaces may reflect all or nearly all of the light falling on the surface. In addition, ordinary, transparent glasses reflect a few percent of the incident light and transmit the rest.

The angle of incidence is i and the angle of reflection i'. Experiment shows that the angles of incidence and reflection are equal, except in a very few peculiar cases, as shown in Fig.1.2.

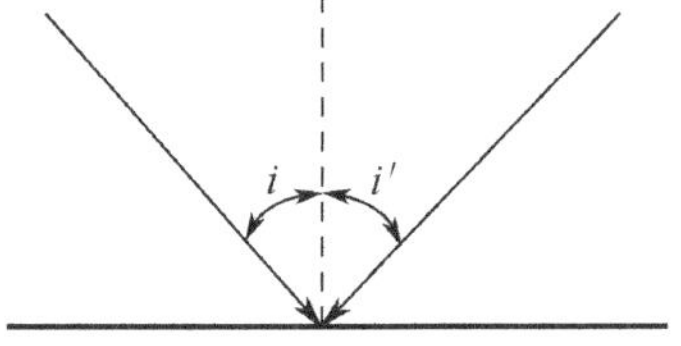

Fig.1.2 Reflection at an interface

We shall later adopt the convention that i is positive; that is, if the acute angle opens counterclockwise from the normal to the ray, i is positive. The sign of i' is clearly opposite to that of i. We therefore write the law of reflection as

$$i' = -i \tag{1.3}$$

1.1.4 Total Internal Reflection

Here we consider a ray that strikes an interface from the high-index side, say, from glass to air (not air to glass). This is known as internal reflection. The law of refraction shows that the incident ray is in this case bent away from the normal when it crosses the interface, as shown in Fig.1.3.

Thus, there will be some angle of incidence for which the refracted ray will travel just parallel to the interface. In this case, $i' = 90°$, so the law of refraction becomes

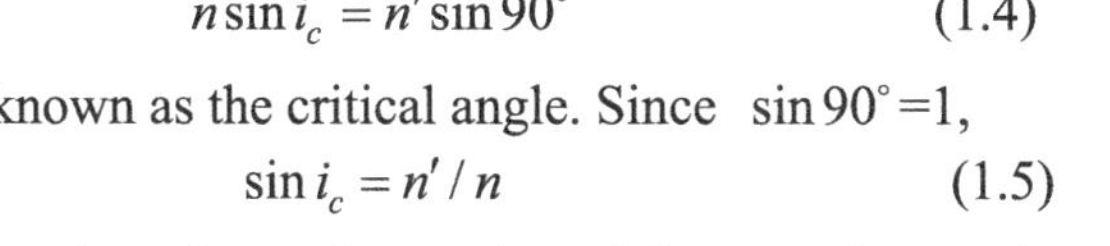

$$n \sin i_c = n' \sin 90° \quad (1.4)$$

where i_c is known as the critical angle. Since $\sin 90° = 1$,

$$\sin i_c = n'/n \quad (1.5)$$

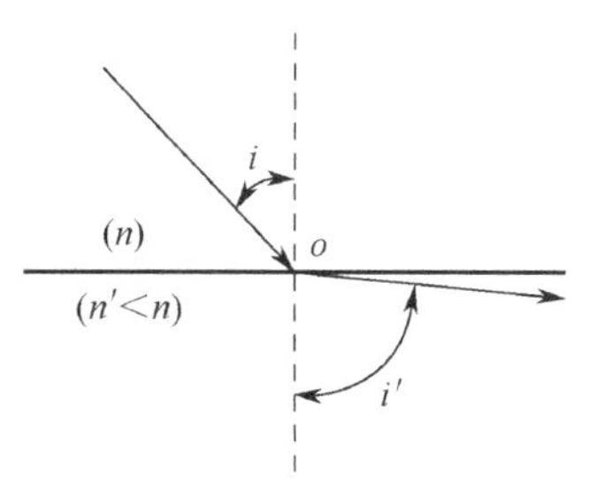

Fig.1.3 Refraction near the critical angle

If i exceeds i_c, then $n \sin i > n'$, and the law of refraction demands that $\sin i'$ exceed 1. Because this is impossible, we can conclude only that there can be no refracted ray in such cases. The light cannot simply vanish, so we are not surprised that it must be wholly reflected; this is indeed the case. The phenomenon is known as total internal reflection; it occurs whenever

$$i > \arcsin(n'/n) \quad (1.6)$$

The reflected light, of course, obeys the law of reflection.

For a typical glass-air interface, n=1.5, the critical angle is about 42°. Glass prisms that exhibit total reflection are therefore commonly used as mirrors with angles of incidence of about 45°.

1.1.5 Reflecting Prisms

There are different types of reflecting prism. The most common are prisms whose cross sections are right isosceles triangles. One advantage of a prism over a metal-coated mirror is that its reflectance is nearly 100% if the surfaces normal to the light are antireflection coated. Further, the prism's properties do not change as the prism ages, whereas metallic mirrors are subject to oxidation and are relatively easy to scratch. A glass prism is sufficiently durable that it can withstand all but the most intense laser beams. Fig.1.4 shows a prism being used in place of a plane mirror.

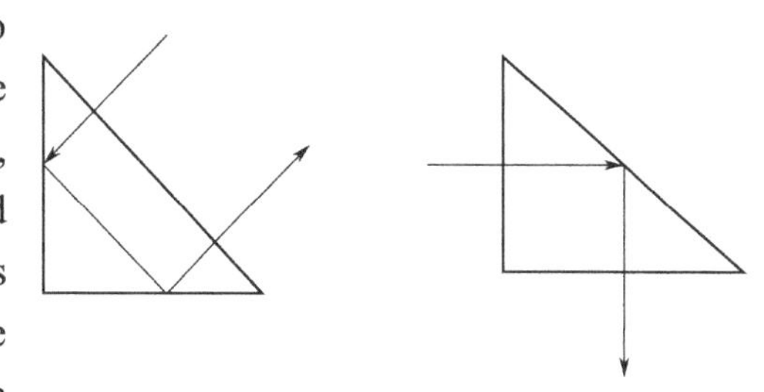

Fig.1.4 Reflecting prisms

In imaging-forming systems, these prisms must be used in collimated light beams to avoid introducing defects into the optical image.

1.2 Imaging

1.2.1 Spherical Surfaces

Because a simple lens consists of a piece of glass with, in general, two spherical surfaces, we will find it necessary to examine some of the properties of a single, spherical refracting surface. We will for brevity call such a surface, as shown in Fig.1.5, a "len". Two of these form a lens. To avoid confusion, we will always place "len" in quotes.

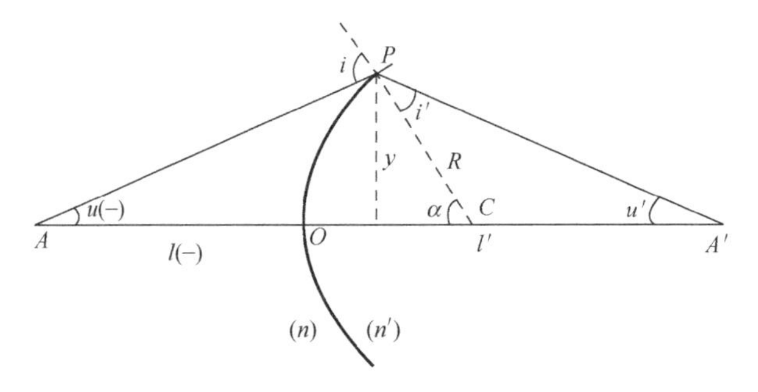

Fig.1.5 Spherical refracting surface

We are interested in the imaging property of the ‘len”. We consider a bright point A and define the axis along the line *AC*, where *C* is the center of the spherical surface. We examine a particular ray *AP* that strikes the “len” at *P*. We shall be interested in the point A' where this ray intersects the axis.

Before proceeding any further, we must adopt a sign convention. The choice of convention is, of course, arbitrary, but once we choose a convention, we shall have to stick with it. The convention we adopt appears, at first, quite complicated. We choose it at least in part because it is universally applicable; with it we will not need to derive a special convention for spherical mirrors.

To begin, imagine a set of Cartesian coordinate axes centered at *O*. Distances are measured from *O*. Distances measured from *O* to the right are positive; those measured from *O* to the left are negative. Thus, for example, OA' and *OC* are positive, whereas *OA* is negative. Similarly, distances measured above the axis are positive; those below are negative. This is our first sign convention.

We now adopt a convention for the signs of angles such as *OAP* or $OA'P$. We determine their signs by trigonometry. For example, the tangent of angle *OAP* is approximately

$$\tan OAP \approx y/OA \tag{1.7}$$

where *y* is the distance indicated between *P* and the axis. Our previous convention shows that *y* is positive, and *OA*, negative. Thus, tan *OAP* is negative and so is *OAP* itself. Similarly, $OA'P$ and *OCP* are positive.

This is our second sign convention. An equivalent statement is that angle $OA'P$ (for example) is positive if it opens clockwise from the axis, or negative otherwise. It is probably simplest, however, merely to remember that angle *OAP* is negative as drawn in Fig.1.5.

Finally, we deal with angles of incidence and refraction, such as angle CPA'. It is most convenient to define CPA' to be positive as shown in Fig.1.5. The angle of incidence or refraction is positive if it opens counterclockwise from the normal (which is, in this case, the radius of the spherical surface）.

Unfortunately, when the last convention is expressed in this way, the statement differs from that which refers to angles (such as *OAP*) formed by a ray crossing the axis. It is best to learn the sign convention by remembering the signs of all of the important angles in Fig.1.5. Only angle *OAP* is negative.

Let us now assign symbols to the more important quantities in Fig.1.5. The point A' is located a distance l' to the right of *O*, and the ray intersects the axis at A' with angle u'. The radius *R* through the point *P* makes angle α with the axis. The angles of incidence and refraction are *i* and i', respectively.

We must be careful of the signs of *OA* and angle *OAP*, both of which are negative according to our sign convention. This is indicated in Fig.1.5 with parenthetical minus signs. We shall later find it necessary, after a derivation based on geometry alone, to go through our formulas and change the signs of all quantities that are algebraically negative. This is so because our sign convention is not used in ordinary geometry. To make our formulas both algebraically and numerically correct, we must introduce our sign convention, which we do as indicated, by changing signs appropriately.

1.2.2 Object-Image Relationship

We now attempt to find a relationship between the quantities l and l' for a given geometry. First, we relate angle u and i to angle α. The three angles in triangle PAC are u, α and $\pi - i$. Because the sum of these angles must be π, we have

$$u+\alpha+(\pi-i)=\pi \tag{1.8}$$

or

$$i=\alpha+u \tag{1.9}$$

Similarly

$$i'=\alpha-u' \tag{1.10}$$

At this point, it is convenient to make the paraxial approximation, namely, the approximation that the ray AP remains sufficiently close to the axis that angles u, u', i and i' are so small that their sines or tangents can be replaced by their arguments; that is

$$\sin\theta=\tan\theta=\theta \tag{1.11}$$

where θ is measured in radians.

It is difficult to draw rays that nearly coincide with the axis, so we redraw Fig.1.5 by expanding the vertical axis a great amount, leaving the horizontal axis intact. The vertical axis has been stretched so much that the surface looks like a plane. In addition, because only one axis has been expanded, all angles are greatly distorted and can be discussed only in terms of their tangents. Thus, for example,

$$u=y/l \tag{1.12}$$

and

$$u'=y/l' \tag{1.13}$$

in paraxial approximation. Note also that large angles are distorted. Although the radius is normal to the surface, it does not look normal in the paraxial approximation.

To return to the problem at hand, the law of refraction is

$$ni=n'i' \tag{1.14}$$

in paraxial approximation, from which we write

$$n(\alpha+u)=n'(\alpha-u') \tag{1.15}$$

Because $OC=R$, we write α as

$$\alpha=y/R \tag{1.16}$$

The last equation therefore becomes

$$n\left(\frac{y}{R}+\frac{y}{l}\right)=n'\left(\frac{y}{R}-\frac{y}{l'}\right) \tag{1.17}$$

A factor of y is common to every term and therefore cancels. We rewrite this relation as

$$\frac{n'}{l'}+\frac{n}{l}=\frac{n'-n}{R} \tag{1.18}$$

At this point, we have made no mention of the sign convention. We derived the proceeding equation on the basis of geometry alone. According to our sign convention, all of the terms in the equation are positive, except l, which is negative. To make the equation algebraically correct, we must, therefore, change the sign of the term containing l. This change alters the equation to

$$\frac{n'}{l'}-\frac{n}{l}=\frac{n'-n}{R} \tag{1.19}$$

which we refer to as the "len" equation.

There is no dependence on y in the "len" equation. Thus, in paraxial approximation, every ray leaving A (and striking the surface) crosses the axis at A'. We therefore refer to A' as the image of A. A and A' are called conjugate points, and the object distance l and image distance l' are called conjugates.

Had we not made the paraxial approximation, the y dependence of the image point would not have vanished. Rays that struck the lens at large values of y would not cross the axis precisely at A'. The dependence on y is relatively small, so we would still refer to A' as the image point. We say that the image suffers from aberrations if all of the geometrical rays do not cross the axis within a specified distance of A'.

1.2.3 Use of the Sign Conventions

A word of warning with regard to the signs in algebraic expression: Because of the sign convention adopted here, derivations based solely on geometry will not necessarily result in the correct sign for a given term. There are two ways to correct this defect. The first, to carry a minus sign before the symbol of each negative quantity, is too cumbersome and confusing for general use. Thus, we adopt the second, which is to go through the final formula and change the sign of each negative quantity. This procedure has already been adopted in connection with the "len" equation and is necessary, as noted, to make the formula algebraically correct. It is important, though, not to change the signs until the final step, lest some signs be altered twice.

1.2.4 Lens Equation

A thin lens consists merely of two successive spherical refracting surfaces with a very small separation between them. Fig.1.6 shows a thin lens in air. The index of the lens is n. The two refracting surfaces have radii R_1 and R_2, both of which are drawn positive.

We can derive an equation that relates the object distance l and the image distance l' by considering the behavior of the two surfaces separately. The first surface alone would project an image of point A to a point A_1'. If A_1' is located at a distance l_1' to the right of the first surface, the "len" equation shows that, in paraxial approximation,

$$\frac{n}{l_1'}-\frac{1}{l}=\frac{n-1}{R_1} \tag{1.20}$$

because n is the index of the glass (second medium) and 1, the index of the air.

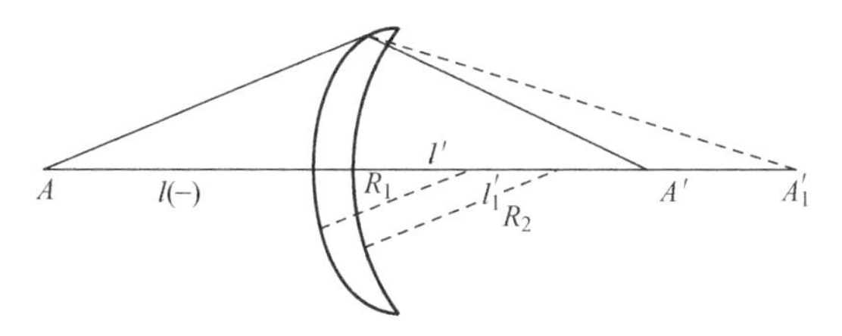

Fig.1.6 Thin lens

The ray does not ever reach A_1', because it is intercepted by the second surface. The second surface, however, behaves as if an object were located at A_1'. The object distance is l_1', if we neglect the thickness of the lens. In applying the "len" equation to the second surface, we must realize that the ray travels across the interface from glass to air. Thus n is the index of the first medium and 1, that of the second. The final image point A' is also the image projected by the lens as a whole. If we call the corresponding image distance l', then the "len" equation yields

$$\frac{1}{l'}-\frac{n}{l_1'}=\frac{1-n}{R_2} \tag{1.21}$$

for the second surface.

If we add the last two equations algebraically, we find that

$$\frac{1}{l'}-\frac{1}{l}=(n-1)\left(\frac{1}{R_1}-\frac{1}{R_2}\right) \tag{1.22}$$

which is known as the lens-maker's formula. The lens-maker's formula was derived from the "len" equation by algebra alone. There are no signs to change because that step was included in the derivation of the "len" equation.

We may define a quantity f' whose reciprocal is equal to the right-hand side of the lens-maker's formula,

$$\frac{1}{f'}=(n-1)\left(\frac{1}{R_1}-\frac{1}{R_2}\right) \tag{1.23}$$

The lens-maker's formula may then be written as

$$\frac{1}{l'}-\frac{1}{l}=\frac{1}{f'} \tag{1.24}$$

where f' is the focal length of the lens. We call this equation the lens equation.

We may see the significance of f' in the following way. If the object is infinitely distant from the lens, then $l=-\infty$. The lens equation then shows that the image distance is equal to f'. If the object is located along the axis of the lens, the image also falls on the axis. We call the image point in this case the secondary focal point F'. Note that any ray that travels parallel to the axis is directed by the lens through F', an observation that we will later find particularly useful.

We define the primary focal point F in a similar way. The primary focal length f is the object distance for which $l'=\infty$. Thus, the lens equation shows that

$$f'=-f \tag{1.25}$$

the primary and secondary focal lengths have equal magnitudes. Any ray that passes through F will be directed by the lens parallel to the axis.

Finally, we note that, in the general case, a lens may have different media on opposite sides. In this case, the lens equation may be shown to be

$$\frac{n'}{l'}-\frac{n}{l}=\frac{n'}{f'}=-\frac{n}{f} \tag{1.26}$$

where n and n' are the indices in the first and second media, respectively. The primary and secondary focal lengths are not equal, but are related by

$$\frac{f'}{f}=-\frac{n'}{n} \tag{1.27}$$

1.2.5 Classification of Lenses and Images

A positive lens is a lens that will cause a bundle of parallel rays to converge to a point. Its secondary focal point lies to the right of the lens and f' is therefore positive. It may be regarded as a lens that is capable of projecting an image of a relatively distant object on a screen. An image that can be projected on a screen is called a real image. In general, a positive lens projects a real, inverted image of any object located to the left of its primary focal point F. When an object is located at F, the image is projected to ∞. The lens is not strong enough to project an image when the object is inside F. In that case, an erect image appears to lie behind the lens and is known as a virtual image.

A positive lens need not have two convex surfaces. It may have the meniscus shape of Fig.1.6. If the lens is thickest in the middle, the lens-maker's formula will show it to be a positive lens.

A negative lens has its secondary focal point located to the left. Its secondary focal length f' is negative, and it cannot project a real image of a real object. Rather, it displays an erect, virtual image of such an object. In only one instance can a negative lens display a real image. This is the case when a positive lens projects a real image that is intercepted by a negative lens located to the left of the image plane. Because the rays are cut off by the negative lens, the real image never appears, but behaves as a virtual object projected by the negative lens.

Like a positive lens, a negative lens need not be concave on both surfaces, but may be a meniscus. If the lens is thinnest in the center, f' will prove to be negative and the lens, also negative.

1.2.6 Spherical Mirrors

Our formalism allows mirror optics to be developed as a special case of lens optics. We notice first that the law of reflection $i'=-i$ can also be written

$$(-1)\sin i' = 1\sin i \tag{1.28}$$

which is precisely analogous to the law of refraction, with $n'=-1$. We may therefore regard a mirror as a single refracting surface, across which the index changes from +1 to −1. It is left as a problem to apply the "len" equation to this case. We find that the focal length of a mirror is

$$f' = R/2 \tag{1.29}$$

where R is the radius of curvature. In addition, the focal points F and F' coincide. The formula that relates the conjugates for a curved-mirror system is

$$\frac{1}{l'}+\frac{1}{l}=\frac{2}{R} \tag{1.30}$$

Mirrors are usually classified as concave and convex. A concave mirror usually projects a real, inverted image, whereas a convex mirror forms an erect, virtual image.

1.2.7 Aberrations

The aberrations of simple, single-element lenses can be quite severe when the lens is comparatively large (with respect to image or object distance) or when the object is located far from the lens axis. When a simple lens is incapable of performing a certain task, it will be necessary to employ a lens, such as a camera lens, whose aberrations have been largely corrected. For specially demanding functions, special lenses may have to be designed and built.

All real lenses made from spherical surfaces may display spherical aberration. Additionally, if the object point is distant from the axis of the lens, or off-axis, the image may display other aberrations, such as astigmatism, coma, distortion, and field curvature. Furthermore, the index of refraction of the lens is a function of wavelength, so its focal length varies slightly with wavelength; the resulting aberration is called chromatic aberration.

Spherical aberration appears both on the axis and off the axis, and does not depend on the distance off-axis. Astigmatism occurs because an off-axis bundle of rays strikes the lens asymmetrically. This asymmetry causes a pair of line images to appear: one behind the plane of best focus and the other in front of it. Coma gives rise to a cometlike image; the head of the comet is the paraxial image point, and the aberration manifests itself as the tail. The tail points away from the axis of the lens and is 3 times longer than its wide. The length of the comatic image, from the paraxial image point to the end of the tail, increases in proportion to the square of the lens diameter and to the distance of the image point from the axis of the lens. The image projected by a lens does not truly lie on a plane but rather on a curved surface, even if other aberrations are zero. This aberration is called field curvature. If the magnification is function of the distance of an image point from the axis, then the image will not be rectilinear. The resulting aberration is called distortion.

Aberrations may be reduced by adjust the radii of the curvature of lens elements so that, for example, angles of incidence are minimized; this process is sometimes called bending the lens. Astigmatism, however, is only weakly influenced by bending the elements. Similarly, one aberration can sometimes be balanced against another. For example, spherical aberration can be partially compensated by moving the image plane from the paraxial image plane to the waist, that is, by compensating spherical aberration by defocusing. Similarly, coma, distortion, and astigmatism can be reduced by adjusting the axis position of the aperture stop.

Words and Expressions

a bundle of	一束
aberration	像差
acute	（尖）锐的，锐角的
algebra /algebraically	代数/用代数的方法
all but	几乎

antireflection	减反射，增透
applicable	可适用的，能应用的，合适的，适当的
approach	方法，路径
arbitrarily	人为地
astigmatism	像散
asymmetry	非对称性
be analogous to	与……类似
be subject to	常遭受
Cartesian coordinate	笛卡儿坐标系
capable	能干的，能胜任的
chromatic aberration	色差
clockwise	顺时针方向的
collimated	准直
coma	彗差
cometlike	彗星状的
concave	凹的
conjugate points	共轭点
convention	习惯，公约，协定
convex	凸的
counterclockwise	逆时针方向的
critical angle	临界角
cross section	横截面
cumbersome	麻烦的，不方便的
defect	缺点，缺陷，瑕疵，损伤
derivations	引出
dispersion	色散
dissimilar	不相似的
distortion	畸变
durable	耐久用的，经久的，坚固的
erect image	正立像
exhibit	呈现，陈列，展出
field curvature	场曲
formalism	体系
geometry	几何学
give rise to	引起，产生，导致
go through	通过
in connection with	与……有关，关于
incident	入射的
inclination	倾斜（角），偏角，倾向
index of refraction	折射率

indices	index 的复数
infinitely	无穷，无限
intact	完整的，原封不动的
intense	强烈的
intercept	截取，拦截，相交，折射
intersect	相交，贯穿
invert image	倒立的像
isosceles	等腰
lest	以免
magnification	放大率
magnitude	数量级
medium	介质
meniscus	弯月形
metallic	金属(制)的
numerically	在数值上
oxidation	氧化
parallel	平行的，类似的
parameter	参数
parenthetical	括号中的
parenthetical	附加的
peculiar	特殊的
polish	抛光，擦亮
primary focal point	主焦点
prism	棱镜
propagate	传播
property	特性，特征
quote	引号
radius	半径
rather	相反地，反而，倒不如说（在句首，或作插入语，其前通常是否定句）
ratio	比，比值
real image	实像
reciprocal	倒数
rectilinear	直线运动的
reflection	反射
refraction	折射
respectively	分别地，各自地
scratch	刮伤，擦伤
secondary focal point	副焦点
sharpness	锐度

single-element lenses	简单透镜
spectrum	谱，光谱
spherical aberration	球差
subscript	下标记
successive	连续的，逐次的，递次的
aperture stop	孔径光阑
paraxial approximation	傍轴近似
tangent	切线
transparent	透明的，半透明的
trigonometry	三角法
universally	一般地，普遍地
vacuo	（拉丁语）真空
vanish	消失，消散
virtual image	虚像
with regard to	关于，论及，对于，就……而言
withstand	抵抗，经得起，经受住

Grammar 专业英语翻译方法（一）：英汉句法对比的总结

英语：

1. 主、谓结构严明，动作行为都有主语；
2. 多被动语态；
3. 介词繁多，名词亦多，应用广泛；
4. 复合句多用“形合法”，根据主、宾、定、状等语法和句法的关系与短语组合成句子，繁而不乱；
5. 语序灵活多变，纵横交错，重点突出，结构严谨，主次分明；
6. 属综合性语言，着重词形、人称、时态、语态、语气的变化；
7. 句法结构复杂，多长句。

汉语：

1. 主、谓结构往往不全，常见无主语句或无人称句；
2. 很少用被动语态；
3. 多用动词，少用介词；
4. 复合句多用“意合法”，句子成分很少用连词，根据事理演变或发展过程和逻辑关系，靠语意串联，承上启下，一气呵成；
5. 语序自然，相映成趣，结构灵活，词句简洁，有如修竹，节节有序；
6. 属分析性语言，根本无词形变化，重意义而不重形态；
7. 句法自然，注意修辞，多简单句型。

Part 2 Wave Optics

In this part we discuss certain optical phenomena for which geometric or ray optics is insufficient. Primarily interference and diffraction, these phenomena arise because of the wave natural of light and often cause sharp departures from the rectilinear propagation assumed by geometric optics. For one thing, diffraction is responsible for limiting the theoretical resolution of a lens to a finite value. This is incomprehensible on the basis of ray optics.

The necessary background for this part is derived with as few assumptions as possible; some familiarity with wave motion is helpful. We cover mainly interference of light, far-field diffraction and just enough near-field diffraction to allow a clear understanding of holography. We shall later devote space to the important interferometric instruments and treat multiple-reflection interference in sufficient detail to apply to laser resonators.

2.1 Waves

2.1.1 Description of Waves

The simplest waves are described by trigonometric functions such as sines, cosines or complex exponential functions. A traveling wave on a string may be described by the equation

$$y = a\cos\frac{2\pi}{\lambda}(x - vt) \tag{2.1}$$

where x is the position along the string and y, the displacement of the string from equilibrium. a is the amplitude of the wave. The wave has crests when the cosine is 1; examining the string at t=0, we can easily see that λ is the distance between crests, or the wavelength.

The argument of the cosine is known as the phase of the wave. Suppose that we consider a constant value Φ_0 of the phase,

$$\frac{2\pi}{\lambda}(x - vt) = \Phi_0 \tag{2.2}$$

and differentiate both sides of this equation with respect to t. The result is

$$v = \mathrm{d}x / \mathrm{d}t \tag{2.3}$$

v is known as the phase velocity, and is the velocity with which a point of constant phase (a crest, for example) propagates along the string. Electromagnetic waves are two dimensional, and v is the propagation velocity of a plane of constant phase in that case.

By examining the wave at one point, for convenience x=0, we may find the frequency ν of the wave. The period τ is the time required for the argument to change by 2π, and ν is the number of periods per second, or the reciprocal of the period. Thus,

$$\nu = v / \lambda \tag{2.4}$$

This relation is more commonly written as

$$\nu\lambda = v \tag{2.5}$$

For compactness, we define two new quantities, wavenumber k,

$$k = 2\pi / \lambda \tag{2.6}$$

and angular frequency ω

$$\omega = 2\pi\nu \tag{2.7}$$

The equation for the wave is then

$$y = a\cos(kx - \omega t) \tag{2.8}$$

Finally, the wave need not have its maximum amplitude when t=0 nor at x=0. We account for this by writing

$$y = a\cos(kx - \omega t + \Phi) \tag{2.9}$$

where Φ is a constant known as relative phase of the wave.

2.1.2 Electromagnetic Waves

Light is a transverse, electromagnetic wave characterized by time-varying electric and magnetic fields. The fields propagate hand in hand; it is usually sufficient to consider either one and ignore the other. It is conventional to retain the electric field, largely because its interaction with matter is in most cases far stronger than that of the magnetic field.

A transverse wave, like the wave on a plucked string, vibrates at right angles to the direction of propagation. Such a wave must be described with vector notation, because its vibration has a specific direction associated with it. For example, the wave may vibrate horizontally, vertically, or in any other direction; or it may vibrate in a complicated combination of horizontal and vertical oscillations. Such effects are called polarization effects. A wave that vibrates in a single plane (horizontal, for example) is said to be plane polarized.

Fortunately, it is not generally necessary to retain the vector nature of the field unless polarization effects are specifically known to be important. This is not the case with most studies of diffraction or interference. Thus, we will generally be able to describe light waves with the scalar equation,

$$E(x,t) = A\cos(kx - \omega t + \Phi) \tag{2.10}$$

where $E(x,t)$ is the electric field strength, A the amplitude and x the direction of propagation.

The speed of light is almost exactly

$$c = 3.00 \times 10^8 \mathrm{m/s} \tag{2.11}$$

and the average wavelength of visible light

$$\lambda = 0.55 \mu\mathrm{m} \tag{2.12}$$

Because $\nu\lambda = c$, the frequency of visible light is approximately

$$\nu = 6 \times 10^{14} \mathrm{Hz} \tag{2.13}$$

Detectors that are able to respond directly to electric fields at these frequencies do not exist. There are detectors that respond to radiant power, however; these are known as square-law detectors. For these detectors, the important quantity is not field amplitude A, but its square, intensity I,

$$I = A^2 \tag{2.14}$$

We should properly use irradiance, but intensity is still conventional where only relative values are required.

2.1.3 Complex Exponential Functions

It is much more convenient to employ complex exponential functions, rather than the trigonometric functions and their cumbersome formulas. For our purposes, complex exponential functions are defined by the relation

$$\exp(\mathrm{i}\alpha) = \cos\alpha + \mathrm{i}\sin\alpha \tag{2.15}$$

where $\mathrm{i}=\sqrt{-1}$. The complex conjugate is found by replacing i with –i.

The electric field is written

$$E(x,t) = A\exp[-\mathrm{i}(kx - \omega t + \Phi)] \tag{2.16}$$

where it is understood that only the real part of E represents the physical wave.

The intensity is defined as the absolute square of the field

$$I(x,t) = E^*(x,t)E(x,t) \tag{2.17}$$

where * denotes complex conjugation. $I(x,t)$ is always real. In a medium whose refractive index is equal to n, the intensity is

$$I = nE^*E \tag{2.18}$$

according to a result of electromagnetic theory.

2.2 Superposition of Waves

Consider two waves, derived from the same source, but characterized by phase difference Φ. They may be written as

$$E_1 = A\mathrm{e}^{-\mathrm{i}(kx-\omega t)} \tag{2.19}$$

and

$$E_2 = A\mathrm{e}^{-\mathrm{i}(kx-\omega t+\Phi)} \tag{2.20}$$

For convenience, we allow them to have the same amplitude. If the waves are superposed, the resultant electric field is

$$E = A\mathrm{e}^{-\mathrm{i}(kx-\omega t)}(1+\mathrm{e}^{-\mathrm{i}\Phi}) \tag{2.21}$$

Before calculating the intensity, we rewrite $(1+\mathrm{e}^{-\mathrm{i}\Phi})$ by removing a factor of $\mathrm{e}^{-\mathrm{i}\Phi/2}$ from both terms

$$(1+\mathrm{e}^{-\mathrm{i}\Phi}) = \mathrm{e}^{-\mathrm{i}\Phi/2}(\mathrm{e}^{-\mathrm{i}\Phi/2} + \mathrm{e}^{\mathrm{i}\Phi/2}) \tag{2.22}$$

or

$$(1+\mathrm{e}^{-\mathrm{i}\Phi}) = 2\mathrm{e}^{-\mathrm{i}\Phi/2}\cos(\Phi/2) \tag{2.23}$$

This technique allows us to write E as a product of real and complex exponential functions only. Because

$$\mathrm{e}^{\mathrm{i}\alpha}\cdot\mathrm{e}^{-\mathrm{i}\alpha} = 1 \tag{2.24}$$

we may immediately write

$$I = 4A^2\cos^2(\Phi/2) \tag{2.25}$$

where A^2 is the intensity of each beam, separately. We shall use this result several times to describe $\cos^2$ fringes.

The intensity of the superposed beams may vary between 0 and twice the sum $2A^2$ of the intensities of the individual beams. The exact value at any point in space or time depends on the relative phase Φ. In particular,

$$I = 4A^2 \quad \text{when} \quad \Phi = 2m\pi \tag{2.26}$$

and

$$I = 0 \quad \text{when} \quad \Phi = (2m+1)\pi \tag{2.27}$$

where m is any integer or zero.

Because energy must be conserved, we realized that we cannot achieve constructive interference without finding destructive interference elsewhere. Interference of two uniform waves may therefore bring about quite complicated distributions of energy.

2.3 Diffraction

Although the distinction is sometimes blurry, we shall say that diffraction occurs when light interacts with a single aperture and that interference occurs when light interacts with several apertures. Diffraction is observed whenever a beam of light is restricted by an opening or by a sharp edge. Diffraction is very often important even when the opening is many orders of magnitude larger than the wavelength of light. However, diffraction is most noticeable when the opening is only somewhat larger than the wavelength.

2.3.1 Huygens' Principle

We can account for diffraction, or at least rationalize its existence, by Huygens' construction. Today, we interpret Huygens' construction as a statement that each point (or infinitesimal area) on a propagating wavefront itself radiates a small spherical wavelet. The wavelets propagate a short (really, infinitesimal) distance, and their resultant gives rise to a "new" wavefront. The new wavefront represents merely the position of the original wavefront after it has propagated a short distance.

More specifically, Huygens' construction is shown in Fig.2.1. The wavefront in this case is a part of a plane wave that has just been allowed to pass through an aperture. A few points are shown radiating spherical wavelets. Both experience and electromagnetic theory indicate that the wavelets are radiated primarily in the direction of propagation. They are thus shown as semicircles rather than full circles.

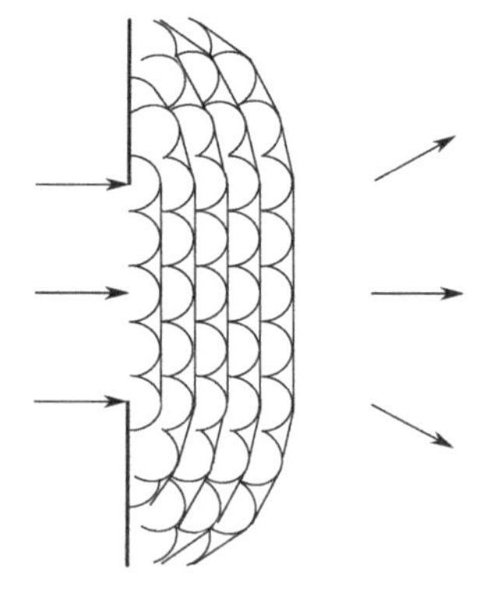
Fig.2.1 Huygens' construction

The spherical wavelets combine to produce a wavefront lying along their common tangent. The new wavefront is nearly plane and nearly identical with the original wavefront. At the edges, however, it develops some curvature owing to the radiation of the end points away from the axis.

Succeeding wavefronts take on more and more curvature, as shown, and eventually the wavefront becomes spherical. We then speak of a diverging wave.

Double-slit interference occurs because diffraction allows the light from the individual slits to interact. Close to slits, where diffraction is not always noticeable, interference is not observed. Only the geometrical shadow of the slits is seen. Far enough from the slits, when the divergence due to diffraction is appreciable, the diffracted beams begin to overlap. From this point on, interference effects are important.

Sufficiently far from the diffracting aperture, we can assume that the rays from the two slits to the point of observation are parallel. This is the simplest case, known as Fraunhofer diffraction or far-field diffraction.

For most diffracting screens, the observing plane would have to be prohibitively distant to allow observation of Fraunhofer diffraction. The approximation is in fact precise only at an infinite distance from the diffracting screen. Fraunhofer diffraction is nevertheless the important case. This is so because the far-field approximation applies in the focal plane of a lens. One way to see this is to recognize that the diffraction pattern, in effect, lies at infinity. A lens projects an image of that pattern into its focal plane.

Finally, we have been tacitly assuming that the diffracting screen is illuminated with plane waves. If this is not so, and it is illuminated with spherical waves originating from a nearby point source, the pattern at infinity is not a Fraunhofer pattern. It is nevertheless possible to observe Fraunhofer diffraction with a well-corrected lens; it can be shown that the Fraunhofer pattern lies in the plane in which the lens projects the image of the point source, no matter what the location of the source. Illuminating with collimated light is just a special case.

2.3.2 Single-Slit Diffraction

This is shown in one dimension in Fig.2.2. We appeal to Huygens' construction and assume that each element ds of the slit radiates a spherical wavelet. The observing screen is located a distance L away from the aperture, and we seek the intensity of the light diffracted at angle θ to the axis.

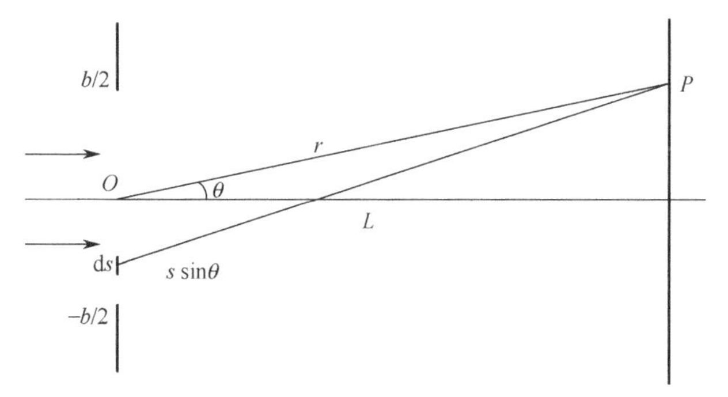

Fig.2.2 Fraunhofer diffraction by a single opening

The center O of the aperture is located a distance r from the observation point P. The optical path difference (OPD) between the paths from θ and from the element ds (at s) is $s\sin\theta$, in

Fraunhofer approximation.

The electric field at P arising from the element is

$$dE = A\frac{e^{-ik(r+s\sin\theta)}}{r}ds \tag{2.28}$$

Here, A is the amplitude of the incident wave, assumed constant across the aperture. We obtain the r in the denominator by realizing that the element is essentially a point source. The intensity from the point source obeys the inverse-square law, so the amplitude falls off as $1/r$. We drop $s\sin\theta$ from the denominator because it is small compared with r. We cannot, however, drop it from the phase term $k(r+s\sin\theta)$ because very small changes of $s\sin\theta$ cause pronounced changes of the phase of the wave relative to that of another wave.

The total field at P is the sum of the fields due to individual elements. If the dimension of the slit is b and its center, $s=0$, this is just the integral

$$E(\theta) = A\frac{e^{-ikr}}{r}\int_{-b/2}^{b/2} e^{-(ik\sin\theta)s}ds \tag{2.29}$$

where constant terms have been removed from the integral. The integrand is of the form $\exp(as)$, so the integral is easily evaluated

$$E(\theta) = A\frac{e^{-ikr}}{r}\frac{2\sin[(kb\sin\theta)/2]}{ik\sin\theta} \tag{2.30}$$

If we multiply both numerator and denominator by b, and define

$$\beta = (kb\sin\theta)/2 \tag{2.31}$$

we may write

$$I(\theta) = \frac{I_0 b^2}{r^2}\left(\frac{\sin\beta}{\beta}\right)^2 \tag{2.32}$$

More-proper analysis, based on electromagnetic theory and a two-dimensional formulation would include an additional factor of $i\lambda$ in the expression for $E(\theta)$, but the important part is the variable $\sin\beta/\beta$.

If the viewing screen is the focal plane of a lens, then the first minimum is located a distance

$$R = \lambda f'/b \tag{2.33}$$

from the center of the pattern, which extends in the direction perpendicular to the edges of the aperture. Over 80% of the diffracted light falls within $2\lambda f'/b$ of the center of the pattern, and the first secondary maximum is about 5% as intense as the principal maximum.

Similar analysis can be carried out with a circular aperture in two-dimensions. The results is similar, except that the pattern is a disk, known as the Airy disk, with radius defined by the first zero as

$$R = 1.22\lambda f'/D \tag{2.34}$$

where D is the diameter of the aperture. It is the finite size of the Airy disk that limits the theoretical resolving power of any optical system.

2.3.3 Fresnel Diffraction

Fresnel diffraction by an aperture refers to the general case, in which either the aperture is not illuminated by a collimated beam or the diffracting screen is not distant compared with the size of the aperture. Diffraction by a single straight edge is always Fresnel diffraction.

2.3.4 Far and Near Field

We have defined Fraunhofer or far-field diffraction as that which is observed whenever the source and observation plane are very distant from diffracting screen. If the diffracting screen is close to either the source or the observation plane, Fresnel or near-field diffraction may be observed. We are now in a position to distinguish more precisely between these two cases.

Consider a diffracting screen whose greatest overall dimension is $2s$. If the source is located at ∞, then the diffracting screen will fall within a single Fresnel zone when the observation point is located a distance s^2/λ beyond the screen. If the observation point is moved closer than s^2/λ the screen will occupy more than one Fresnel zone. This is known as the region of Fresnel diffraction.

In the same way, if the observation point is moved beyond s^2/λ, the screen occupies less than one Fresnel zone. This is the region of Fraunhofer diffraction. In this region, all points on the screen are equidistant from the observation point, to an accuracy of better than $\lambda/4$. This corresponds closely to our assumptions when we derived the diffraction pattern of a slit.

The simple pinhole camera can be used to illustrate near- and far-field diffraction. The pinhole camera is a small hole punched in an opaque screen, with a viewing screen located a distance f' beyond the hole. When the hole is very large, the image of a distant point is the geometrical shadow of the opaque screen. The diameter of the image is thus equal to the diameter of the pinhole. The limit of resolution in the image plane is about 1.5 times larger. When the hole is small, Fraunhofer diffraction applies, and the limit of resolution is $0.61\,\lambda f'/s$, where s is the radius of the pinhole. For intermediate-size holes, Fresnel diffraction applies, and little can be said without detailed calculation.

2.4 Interference

In this section, we will discuss two types of interference, i.e., interference by division of wavefront and interference by division of amplitude.

2.4.1 Interference by Division of Wavefront

The wavefront refers to the maxima (or other planes of constant phase) as they propagate. The wavefront is normal to the direction of propagation. One way of bringing about interference is by the wavefronts in two or more segments and recombining the segments elsewhere.

Suppose a monochromatic plane wave (a collimated beam, or a beam with plane wavefronts) is incident on the opaque screen, as shown in Fig.2.3. Two infinitesimal slits a distance d apart have been cut into the screen. Each slit behaves as a point source, radiating in all directions. We set up an observing screen a great distance L away from the slits. Light from both slits falls on this screen.

The electric field at a point P is the sum of the fields originating from each slit, i.e.,

$$E = A(e^{-ikr_1} + e^{-ikr_2})e^{i\omega t} \tag{2.35}$$

where A is the amplitude of the waves at the viewing screen and r_1, r_2 are the respective distances of the slits from P. Because the factor $e^{i\omega t}$ is common to all terms and will vanish from the intensity, we shall hereafter drop it.

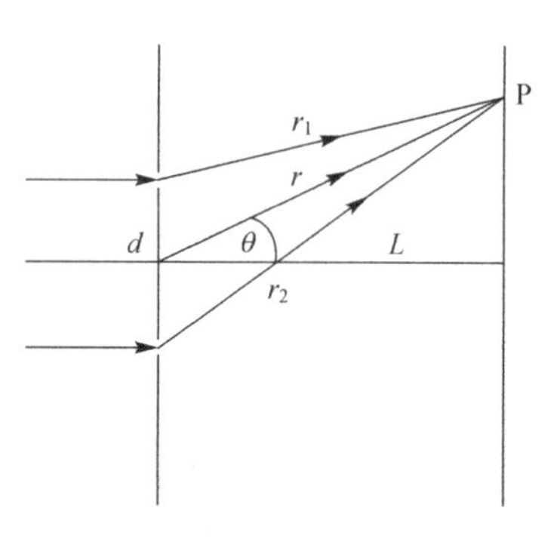

Fig.2.3 Double-slits interference

If L is sufficiently large, r_1 and r_2 are effectively parallel and differ only by optical path difference (OPD) $d\sin\theta$. Thus, we obtain

$$E = Ae^{-ikr_1}(1 + e^{-ikd\sin\theta}) \tag{2.36}$$

and the intensity can be written as

$$I = 4A^2\cos^2\left(\frac{\pi}{\lambda}d\sin\theta\right) \tag{2.37}$$

For small angles, $\sin\theta = x/L$, and the interference pattern has a $\cos^2$ variation with x. Maxima occur when the argument of the cosine is an integral multiple of π, or where

$$\text{OPD} = m\lambda \qquad \text{(constructive interference)} \tag{2.38}$$

Similarly, minima (in this case, zeros) occurs whenever

$$\text{OPD} = (m+1/2)\lambda \qquad \text{(destructive interference)} \tag{2.39}$$

In fact, the $\cos^2$ fringes do not extend infinitely far from the axis. This is so for at least two reasons: (a) the light is not purely monochromatic, and (b) the slits are not infinitesimal in width. The first relates to the coherence of the light. This effect brings about a superposition of many double-slit patterns, one for each wavelength. Each wavelength brings about slightly different fringe pattern from the rest, and at large angles θ the patterns do not coincide exactly. This results in the washing out and eventual disappearance of the fringes. The second effect has to do with diffraction. This assumption is valid only for zero slit width. A finite slit radiates primarily into a cone whose axis is the direction of the incident light. For this reason, the intensity of the pattern falls nearly to 0 for large θ.

If we generalize from two slits to many, we find that the OPD between rays coming from adjacent slits is $d\sin\theta$. Thus, the OPD between the first and the jth slit is $(j-1)\ d\sin\theta$. The total electric field at a point on the distant observation screen is a sum of many terms, i.e.,

$$E = Ae^{-ikr_1}[1 + e^{-i\phi} + e^{-2i\phi} + e^{-3i\phi} + \cdots + e^{-(N-1)i\phi}] \tag{2.40}$$

where N is the number of slits and $\phi = kd\sin\theta$ as before. Thus, the intensity of the interference pattern is

$$I(\theta) = A^2\frac{\sin^2 N\phi/2}{\sin^2\phi/2} = A^2\frac{\sin^2\left(\frac{\pi}{\lambda}Nd\sin\theta\right)}{\sin^2\left(\frac{\pi}{\lambda}d\sin\theta\right)} \tag{2.41}$$

A typical interference pattern is illustrated in Fig.2.4. The sharp peaks are known as principal maxima and appear only when

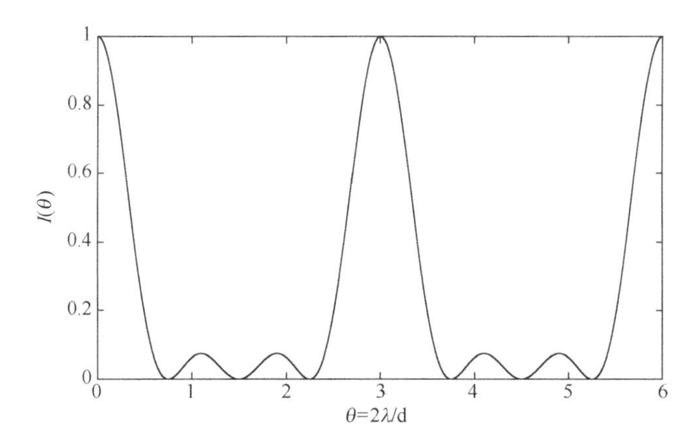

Fig.2.4 Four-slit interference pattern

$$\frac{\pi}{\lambda}d\sin\theta = m\pi\ ; \quad m = 0,\ \pm 1,\ \pm 2, \cdots \tag{2.42}$$

or when

$$m\lambda = d\sin\theta \tag{2.43}$$

This is known as the grating equation, and m is known as the order number or order.

The smaller peaks are called secondary maxima and appear because of the oscillatory nature of the numerator of $I(\theta)$. When N>>1, the secondary maxima are relatively insignificant, and the intensity appears to be 0 at all angles where the grating equation is not satisfied. At all angles that satisfied the grating equation, the intensity is N^2A^2; it falls rapidly to 0 at other angles.

2.4.2 Interference by Division of Amplitude

Interference devices based on division of amplitude use a partial reflector to divide the wavefront into two or more parts. These parts recombined to observe the interference pattern.

Consider two parallel surfaces with relatively high reflectance and observe the effect of multiple reflections, as shown in Fig.2.5.

Assume that the viewing screen is a great distance away. The total field on the view screen can be expressed as

$$E = At^2(1 + r^2\mathrm{e}^{-\mathrm{i}\phi} + r^4\mathrm{e}^{-2\mathrm{i}\phi} + \cdots) \tag{2.44}$$

where t is the amplitude transmittance of the surface, r is the amplitude reflectance, and the phase change between two reflections can be written as

$$\phi = 2kd\cos\theta \tag{2.45}$$

Then, the transmitted intensity is written after some manipulation as

$$I_T = I_0\left(\frac{T}{1-R}\right)^2 \frac{1}{1 + F\sin^2\dfrac{\phi}{2}} \tag{2.46}$$

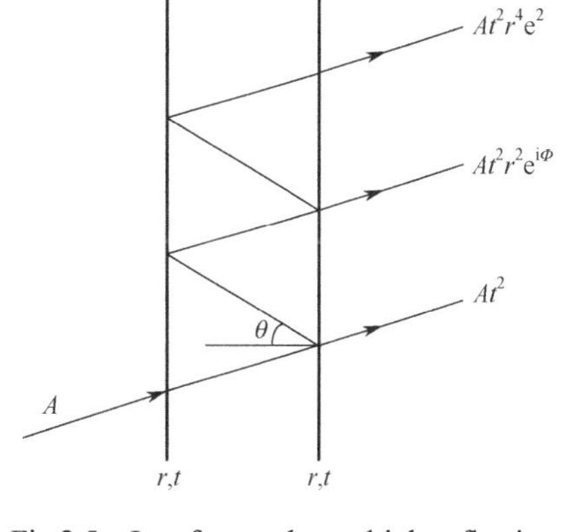

Fig.2.5 Interference by multiple reflections

where $T = t^2$, $R = r^2$, and $F = 4R/(1-R)^2$.

The factor $T/(1-R)$ is a constant, and we shall for convenience drop it hereafter and assume perfect reflectors. Fig.2.6 shows the multiple-reflection interference pattern. Maxima occur when

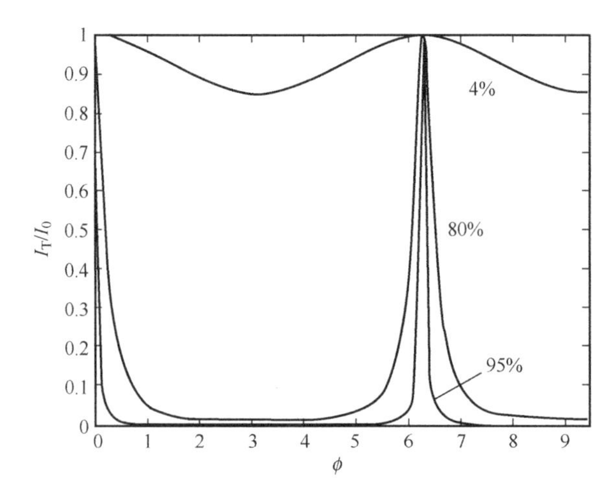

Fig.2.6 Multiple-reflection interference pattern

$$\phi / 2 = m\pi$$

or when

$$m\lambda = 2d\cos\theta, \quad m = 0, 1, 2, \cdots \tag{2.47}$$

For fixed value of d, we will observe transmission at fixed value of θ. If the plates are illuminated with a range of angles, we will see a series of bright rings. On the other hand, if we remain at $\theta = 0$ and vary d, we will observe transmission at specific values of d only.

2.4.3 Michelson Interferometer

In 1881, A. A. Michelson designed an accuracy interference device based on component amplitude interferometry, and the simplest form of the instrument is shown in Fig.2.7. Light from an extended source S is divided at the semireflecting surface A of a plane parallel glass plate D into two beams. These are reflected at plane mirrors M_1, M_2, and return to D, where they are re-combined to enter the observing telescope T. M_2 is fixed, while M_1 is mounted on a carriage and can be moved towards or away from D by means of a micrometer screw. The beam reflected from M_1 traverses the dispersive material of D three times before reaching T, compared with a single passage for the beam reflected from M_2. To remove this asymmetry, a compensating plate C with the same material and thickness of D is introduced between D and M_2, and it is parallel to D.

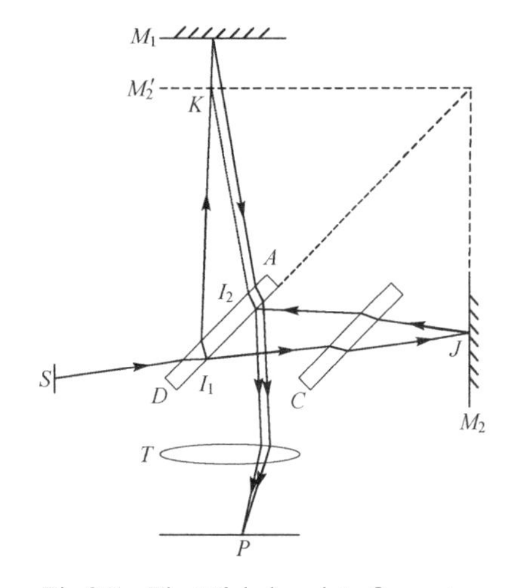

Fig.2.7 The Michelson interferometer.

Suppose M_2' is the image of M_2 in the beam-divider. The optical path between S and the point P along a ray SI_1JI_2P is equal to the optical path between S and P along the ray SI_1KI_2P. The interference pattern observed with the telescope may therefore be considered to arise from an air film bounded by the real reflecting surface M_1 and the virtual reflecting surface M_2', provided we associate with the latter a phase change ϕ equal to the difference between the phase changes for external and internal reflection at A. The value of ϕ depends on the nature of the semi-reflector A.

When M_1 and M_2' are parallel, the fringes given by a quasi-monochromatic source are circle and localized at infinity. If M_1 is moved to approache M_2', the fringes contract towards the center. However, the angular scale of the pattern increase until the illumination over the field of view is uniform at a level when M_1 coincides with M_2'. The mirrors M_1 and M_2 are then said to be in optical contact. When M_1 and M_2' are close together but mutually inclined to form a wedge with small angle, there are fringes localized at or near the surface of this wedge. As the separation increases, however, the range of incidence angle corresponding to each point of the field of view and the variation of mean incidence angle over the field of view can not to be negligible. Thus, the visibility of the fringes decreases, and they become curved with convex side towards the wedge apex. Whether M_1 and M_2' are parallel or inclined, a change $\Delta m\lambda_0$ of the optical path in either arm of the instrument results in a displacement of the pattern through Δm orders.

When the separation of M_1 and M_2' is only a few wavelengths, fringes are visible with white light. They are used to recognize a reference fringe in the monochromatic pattern. If $|\phi| = \pi$, the central fringe of the white light pattern is black and defines the intersection of M_1 and M_2', so that it is in the same position as the monochromatic fringe with $|m| = 1/2$. Otherwise, the achromatic fringe does not in general coincide with a bright or dark fringe of the monochromatic pattern, but this presents no difficulty if the transfer between white light and monochromatic patterns is made consistently.

2.4.4 Fabry-Perot Interferometer

The multiple beam interference fringes from a plane-parallel plate illuminated near normal incidence are used in the Fabry-Perot interferometer. This instrument consists essentially of two glass or quartz plates P_1, P_2 with plane surfaces, as shown in Fig.2.8. The inner surfaces are coated with partially transparent films of high reflectivity and are parallel, so that they enclose a plane-parallel plate of air. The plates themselves are made slightly prismatic in order to avoid disturbing effects due to reflection at the outer uncoated surfaces. In the original form of the instrument, one plate was fixed and the other was mounted on a screw-controlled carriage to allow continuous variation of the plate separation, but this arrangement is obsolete because of difficulties of mechanical construction. Instead, the plates are separated by a fixed spacer D, which is commonly a hollow cylinder of invar or silica with three projecting studs at each end, and the plates are kept in place by the pressure of springs. The spacer is optically worked so that the planes defined by the studs are as nearly parallel as possible, and fine adjustment can be made by varying the spring pressure. This form of the interferometer with fixed

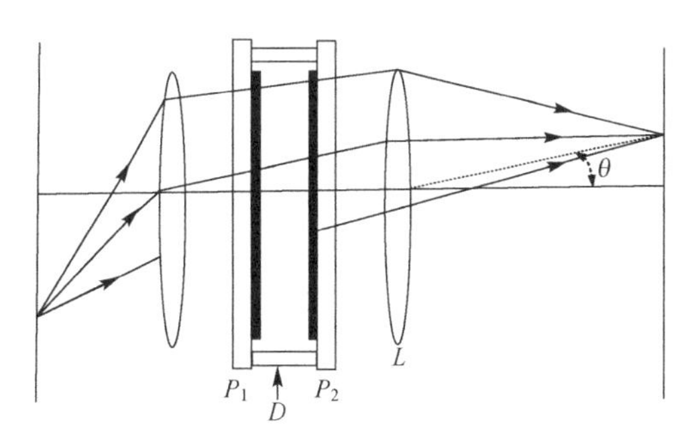

Fig.2.8 The Fabry-Perot interferometer

plate separation is somethings called a Fabry-Perot etalon.

A light from an extended quasi-monochromatic source S forms narrow bright fringes of equal inclination in the focal plane of the lens L. The order of interference is given by

$$m=\frac{\delta}{2\pi}=\frac{2n'h\cos\theta'}{\lambda_0}+\frac{\phi}{\pi} \tag{2.48}$$

where n' is the refraction index of the medium between the plates, h denotes the separation of the reflecting surfaces, θ' is the angle of reflection, and ϕ represents the phase change. The axis of the lens is usually normal to the plates, and the bright fringes corresponding to integral values of m are then circles with common center at the focal point for normally transmitted light (Fig.2.8). At this point, m has maximum value m_0 given by

$$m_0=\frac{2n'h}{\lambda_0}+\frac{\phi}{\pi} \tag{2.49}$$

In general, m_0 is not an integer, and we may write

$$m_0=m_1+e \tag{2.50}$$

where m_1 is the integral order of the innermost bright fringe. e, which is less than unity, is the fractional order at the center. The bright fringes is

$$2n'h\cos\theta'\pm\frac{\lambda}{2}=m\lambda_0,\quad m=0,1,2,\cdots \tag{2.51}$$

Form (2.48) to (2.51), we obtain the angular radius θ_p of the pth bright fringe from the center when θ_p is not too large, i.e.,

$$\theta_p=\frac{1}{n}\sqrt{\frac{n'\lambda_0}{h}}\sqrt{p-1+e} \tag{2.52}$$

where n is the refractive index of the medium outside the plates. The diameter D_p of this fringe is therefore given by

$$D_p^2=(2f\theta_p)^2=\frac{4n'\lambda_0 f^2}{n^2 h}(p-1+e) \tag{2.53}$$

where f is the focal length of the lens L.

2.5 Coherence

Coherence describes the statistical similarity of a field at two points in space or time. Until now, we have almost always assumed light to be completely coherent, in the sense that any interference experiment resulted in high-quality interference fringes. In general, this is not the case, except with certain laser sources; the light from most sources is said to be incoherent or partially coherent. When conditions are such that the light is incoherent, it is not possible to detect interference effects. A discussion of wave optics is incomplete without considering the conditions that must exist for an interference experiment to be performed successfully.

Temporal coherence is the measure of the average correlation between the value of a wave and itself delayed by τ, at any pair of times. Temporal coherence characterizes how well a wave can interfere with itself at a different time. The delay over which the phase or amplitude wanders by a significant amount (and hence the correlation decreases by significant amount) is defined as the coherence time τ_c. At a delay of τ = 0 the degree of coherence is perfect, whereas it drops significantly as the delay passes $\tau = \tau_c$. The coherence length L_c is defined as the distance the wave travels in time τ_c. It can be shown that the larger the range of frequencies Δf a wave contains, the faster the wave decorrelates (and hence the smaller τ_c is). The characteristic time $\tau_c = 1/\Delta f$ is of the order of the coherence time.

Consider next the light disturbance at two points P_1 and P_2 in a wave field produced by an extended quasi-monochromatic source S. For simplicity, assume that the wave field is in vacuum and that P_1 and P_2 are many wavelengths away from the source. We may expect that, when P_1 and P_2 are close enough to each other, the fluctuations of the amplitudes at these points and also the fluctuations of the phases will not be independent. It is reasonable to suppose that, if P_1 and P_2 are so close to each other that the difference $\Delta S = SP_1 - SP_2$ between the paths from each source point is small compared to the mean wavelength $\bar{\lambda}$, then the fluctuations at P_1 and P_2 will effectively be the same; and that some correlation between the fluctuations will exist even for greater separations of P_1 and P_2, provided that for all source points the path difference ΔS does not exceed the coherence length $c\tau_c \sim c/\Delta f = \bar{\lambda}^2/\Delta\lambda$.

Light sources are put into one of two categories, laser sources and thermal sources. A typical thermal source is a gas discharge lamp. In such a lamp, light is emitted by excited atoms that are, in general, unrelated to each other. Each atom emits relatively short bursts or wave packets. If an atom is excited several times, it can emit several consecutive wave packets. These packets are generally far apart (compared with their duration) and are emitted randomly in time. The packets emitted by a single atom therefore bear no constant phase relation with each other.

Suppose we try to perform interference by division of amplitude with the packets emitted by a single atom. The wave reflected from the second surface is delayed with respect to the first because of the finite speed of light. If the delay is greater than the duration of the wave packet, the two reflected packets will not reach the detector simultaneously. There will therefore be no interference pattern, and we would compute the intensity at the viewing screen by adding the intensities (not amplitudes) of the reflected waves. The light is said to be incoherent for the purpose of this experiment.

This statement would be true even if the wave packets were emitted so rapidly that several packets entered the apparatus at the same time. (In this case, the light would undergo rapid amplitude and phase fluctuations.) Because the packets are emitted at random, they bear no definite phase relation. We would sometimes detect a maximum and sometimes a minimum for any given OPD. Over the long term, we would observe constant intensity and would regard the light as incoherent.

Similarly, the waves from one atom bear no definite relation with the waves from any other atom. By precisely the same reasoning, we conclude that the light emitted by one atom is incoherent with that emitted by any other atom. Superposition of the waves from different atoms is therefore described by adding intensities, not amplitudes.

2.6 Polarization

Polarization is a property applying to transverse waves that specifies the geometrical orientation of the oscillations. In a transverse wave, the direction of the oscillation is perpendicular to the direction of motion of the wave. Natural light, and most other common sources of visible light, are incoherent: Radiation is produced independently by a large number of atoms or molecules whose emissions are uncorrelated and generally of random polarizations. In this case, the light is said to be unpolarized. This term is somewhat inexact, since at any instant of time at one location there is a definite direction to the electric and magnetic fields. However, it implies that the polarization changes so quickly in time that it will not be measured or relevant to the outcome of an experiment.

The polarization state of light traveling along the z axis, for example, is described by superimposing two electric fields whose directions are parallel to the x and y axes. In this case, we can express the electromagnetic wave traveling along the z axis as the vector sum of the electric fields E_x and E_y:

$$E(z,t)=\boldsymbol{E}_x(z,t)+\boldsymbol{E}_y(z,t)=\{E_{x0}\exp[\mathrm{i}(\omega t-Kz+\delta_x)]\}\boldsymbol{x}+\{E_{y0}\exp[\mathrm{i}(\omega t-Kz+\delta_y)]\}\boldsymbol{y} \quad (2.54)$$

where $\boldsymbol{x}$ and $\boldsymbol{y}$ are unit vectors along the coordinate axes. When we describe the states of polarization, the absolute values of the initial phases (δ_x, δ_y) are not required, and only the relative phase difference ($\delta_x - \delta_y$) or ($\delta_y - \delta_x$) is taken into account. Similarly, only the phase difference is accounted for in spectroscopic ellipsometry.

Figure 2.9 shows the variation of the polarization state with the phase difference δ_y - δ_x. In this figure, $E_{x0} = E_{y0}$ and $K = 1$ are assumed.

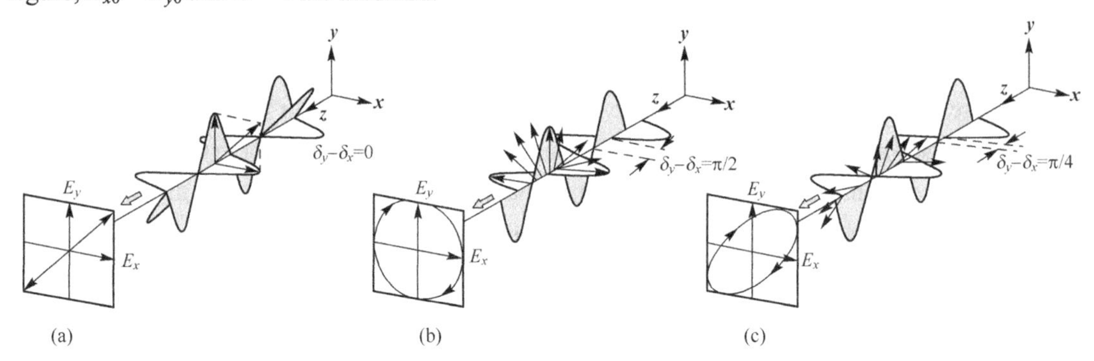

Fig.2.9 Representation of (a) linear polarization, (b) right-circular polarization and (c) eppliptical polarization. Phase differences between the electric fields parallel to the x and y axes ($\delta_y - \delta_x$) are (a) 0, (b) $\pi/2$, and (c) $\pi/4$.

As shown in Fig.2.9, when $\delta_y - \delta_x = 0$, there is no phase difference between E_x and E_y, and the orientation of the synthesized vector ($E_x + E_y$) is always 45° in the x-y plane. In other words, an electromagnetic wave oriented at 45° can be resolved into two electromagnetic waves vibrating parallel to the x and y axes. However, the amplitude of the synethesized vector is $\sqrt{2}$ times larger than that of E_{x0} or E_{y0}. The polarization state shown in Fig.2.9(a) is referred to as linear polarization. When the phase difference between E_x and E_y is 90° ($\delta_y - \delta_x = \pi/2$), the synthesized vector rotates in the x-y plane as the light propagate. This polarization state shown in Fig.2.9(b) is known as circular polarization. As confirmed from Fig.2.9, since $\delta_y - \delta_x = \pi/2$ and $K = 1$ in this case, the wave oscillating along the y axis advances forward by $\pi/2$, compared with the wave oscillating along the x axis. In Fig.2.9(b), if we choose a point on the z axis, the synthesized vector on the point rotates toward the right (clockwise) as the light propagates with time (the rotation is counterclockwise in the positive direction of the z axis). This particular polarization is called right circular polarization. The polarization shown in Fig.2.9(c) is referred to as elliptical polarization and rotates toward the right (clockwise) when $\delta_y - \delta_x = \pi/4$.

Words and Expressions

airy disk	艾里斑
amplitude	振幅
aperture	孔径，光阑
apex	顶点，间断
apparatus	装置，设备，仪器
appeal to	求助于
appreciable	可以察觉到的，可观的
be in a position to	处在可以（做……）的位置，能够
be responsible for	造成，引起，导致，是造成……的主要原因
bear relation to	与……有关系
blurry	模糊的
burst	脉冲，闪光，波群
category	类别
coherence	相干性
collimated	准直的
compactness	简洁
complex conjugate	复共轭
complex exponential function	复指数函数
consecutive	连续的
constant	常数
constructive interference	相长干涉
counterclockwise	反时针方向的
convex	凸的，凸面的
correspond	相符合，相对应
correlation	相关性
crest	峰值
curvature	弯曲的
denominator	分母
departure	偏差，偏移
derive	推导
destructive interference	相消干涉
differentiate	微分，求导
diffraction	衍射

displacement	位移
distinction	区别
diverge	发散
double-slit interference	双缝干涉
duration	持续时间
elliptical	椭圆的
employ	使用
equilibrium	平衡位置
familiarity with	熟悉，精通
fluctuation	涨落，起伏
focal plane	焦平面
for one thing	举例来说
fringe	条纹，边纹，边带
gas discharge lamp	气体放电灯
hand in hand	联合地，共同地
holygraphy	全息术
identical	相同的
in the sense that	从某种意义上说
incoherent	非相干的
incomprehensible	不可思议的
infinitesimal	无穷小
insufficient	不能胜任的
intensity	强度
interaction	相互作用
interference	干涉
interferometer	干涉仪
intermediate	中等的
invar	不胀钢，因瓦合金
inverse-square law	平方反比定律
irradiance	辐照度
laser	激光
monochromatic	单色的
more specifically	更准确地说
motion	运动
mounted	安装
multiple	多重的
nevertheless	尽管如此，仍然
noticeable	显而易见的
numerator	分子
obsolete	过时的，废弃的

occupy	占据
opaque	不透明的
optical path difference	光程差
orders of magnitude	数量级
oscillation	振荡
overlap	重叠
partially coherent	部分相干的
passage	通道
pattern	图案，模式
perpendicular	与……垂直，正交
phase velocity	相速度
pinhole	针孔
polarization	偏振，极化
primarily	主要是，基本上，首要地
prismatic	棱柱的，棱镜的
punch	穿孔，打孔
quartz	石英
quasi-monochromatic	准单色的
radiant power	辐射功率
rationalize	合理地说明
reciprocal	倒数
rectilinear	直线的
resolution	分辨率
resonator	谐振腔
semireflecting	半反射
scalar	标量
screw-controlled carriage	螺旋控制托架
sharp	尖锐的
simultaneously	同时
spectroscopic ellipsometry	椭圆偏振光谱测量
square-law detector	平方律检波器
superposition	叠加
synthesized vector	合成矢量
tacitly	不言而喻地
tangent	切线，正切
thermal sources	热辐射源
theoretical	理论上的
this is not the case	情况并非如此
transverse	横向的
trigonometric functions	三角函数

undergo	经历
valid	有效的
variable	变量
vector	矢量
vibrate	振动
wave packet	波包，波群
wavelet	子波
wedge	楔形物
with respect to	关于，至于

Grammar　专业英语翻译方法（二）：被动语态的译法

专业英语中多用被动语态，而汉语中多用主动句。

1. 被动语态的译法

尽可能译成汉语主动句。

—The piston is forced to move downward by the expanding gas.
膨胀的气体推动活塞下行。

—The pump is being adjusted by the assistant engineer.
助理工程师正在调整泵。

（1）句子结构基本不变，汉译时只变换语态，用“将”“使”“得到”等，译成汉语主动句。

—The plan has been overfulfilled.
计划超额完成了。

—This formula has already been mentioned above.
这公式上面已提过。

—Temperature is changed quickly from room temperature to 125°C and is held there for at least 15 minutes.
使温度很快从室温升高到 125℃，并至少保持 15min。

（2）不变动无生命的主语形式而改变谓语结构。

—The production has been greatly increased.
产量已经有了很大的提高。

—The discovery is highly appreciated in the circle of science.
这一发现在科学界中得到很高的评价。

（3）将主语译成宾语或另加适当的主语。

—We are brought freedom and happiness by socialism.
社会主义给我们带来了自由和幸福。

（4）将主语译成宾语，构成汉语无人称句。

—Temperature is taken at regular times.
定时记录温度。

—The quality of products should be paid attention to.
应该注意产品质量。

—Smoking is not permitted in this theatre.
本剧院内禁止吸烟。

（5）带情态动词的被动结构，汉译时用“可以”或“不可以”。

—That man cannot be relied upon.
那人不可信赖。

（6）译成汉语被动句型，除个别句子用“被”字外，更多一些句子可用“受”“由”“给”“加以”“予以”“为……所……”等。

—In the rains, things have been affected with damp.
在雨季，东西受潮了。

—Many works at present are done by machines.
许多工作现在是由机器完成的。

—This plan will be discussed first.
这个计划将首先予以考虑。

（7）谓语分译，译成带主语或不带主语的独立成分。

—Hydrogen is known to be the lightest element.
人们知道，氢是最轻的元素。

（8）非限定动词构成的被动语态。

A．不定式，与情态动词“have to”构成复合谓语。

—The oxygen has to be removed in the process.
氧必须在此过程中除去。

B．不定式做逻辑主语。

—It is impossible for heat to be converted into a certain energy without something lost.
把热转变成某一种能而不损失一些是不可能的。

C．现在分词

—Being cooled in the air, the metal hardened.
金属在空气中冷却时就硬化了。

D．过去分词

—The force required to move a body depends on its mass and acceleration.
移动一个物体所需的力取决于它的质量和加速度。

E．动名词

—The constructor insists upon his device being tested under operating condition.
设计师坚持他的装置要在工作条件下实验。

2．科技英语中有很多以“it”做被动结构的形式主语

It is considered that	人们认为
It is expected that	预期，人们希望
It has proved that	业已证明
It has been shown that	有人指出

It is believed that	人们相信
It is estimated that	据估计
It is found that	人们发现，人们认为
It is required that	需要
It may be safely said that	可以有把握地说
It must be admitted that	老实说，必须承认
It must be emphasized that	必须强调
It should be pointed that	必须指出
It is said that	据说
It is suggested that	有人建议
It is supposed that	据推测
It is usually considered that	通常认为
It is well-known that	众所周知
It will be seen that	由此可见

Part 3 Holography and Fourier Optics

We begin this part by using simple arguments to describe holography and to treat such important aspects of holography as image position, resolution and change of wavelength. Under the general heading of Optical Processing, we include the Abbe theory of the microscope, spatial filtering, phase-contrast microscopy and matched filtering; in short, what is often called Fourier-transform optics.

3.1 Holography

Holography or wavefront reconstruction was invented before 1948, about twenty years before the development of the laser. Its real success came only after the existence of highly coherent sources suggested the possibility of separating the reference and object beams to allow high-quality reconstructions of any object.

3.1.1 Principle of Holography

In Gabor's original holography, light was filtered and passed through a pinhole to bring about the necessary coherence. The source illuminated a small, semi-transparent object O that allowed most of the light to fall undisturbed on a photographic plate H. In addition, light scattered or diffracted by the object also falls on the plate, where it interferes with the direct beam or coherent background. The resulting interference pattern may be recorded on the plate and contains enough information to provide a complete reconstruction of the object.

To find the intensity at H we may write the field arriving at H as

$$E = E_i + E_0 \tag{3.1}$$

where E_i is the field due to the coherent background and E_0, the field scattered from the object. The scattered field E_0 falling on H is not simple, both amplitude and phase vary greatly with position. We therefore write

$$E_0 = A_0 \mathrm{e}^{\mathrm{i}\psi_0} \tag{3.2}$$

where A_0 and ψ_0 are implicitly functions of position. We write a similar expression for E_i, even though E_i is usually just a spherical wave with nearly constant amplitude A_i. The field falling on the plate may then be expressed as

$$E = \mathrm{e}^{\mathrm{i}\psi_i}[A_i + A_0 \mathrm{e}^{\mathrm{i}(\psi_0 - \psi_i)}] \tag{3.3}$$

and the intensity

$$I = A_i^2 + A_0^2 + A_0 A_i \mathrm{e}^{\mathrm{i}(\psi_0 - \psi_i)} + A_0 A_i \mathrm{e}^{-\mathrm{i}(\psi_0 - \psi_i)} \tag{3.4}$$

For our purpose it is convenient to characterize the photographic plate H by a curve of amplitude transmittance t_a vs exposure ζ rather than by the D vs $\log\zeta$ curve of conventional photography. The curve is nearly linear over a short region; we call the slope there β.

In that region, the equation for the t_a vs ζ curve can be written as

$$t_a = t_0 - \beta\zeta \tag{3.5}$$

for the (negative) emulsion.

If we take the exposure time to be t, we find the amplitude transmittance of the developed plate to be

$$t_a = t_0 - \beta t[A_i^2 + A_0^2 + A_0 A_i \mathrm{e}^{\mathrm{i}(\Psi_0 - \Psi_i)} + A_0 A_i \mathrm{e}^{-\mathrm{i}(\Psi_0 - \Psi_i)}] \tag{3.6}$$

The term A_0^2 contributes to noise in the reconstruction, but we drop it here because of our assumption that the scattered field is small compared with the coherent background.

We may now remove the object and illuminate the developed plate with the original reference beam E_i. The developed plate is known as the hologram. The transmitted field E_t just beyond the hologram is

$$E_t = A_i \mathrm{e}^{\mathrm{i}\Psi_i} t_a \tag{3.7}$$

and the interesting part is

$$-\beta t A_i \mathrm{e}^{\mathrm{i}\Psi_i}[A_i^2 + A_0 A_i \mathrm{e}^{\mathrm{i}(\Psi_0 - \Psi_i)} + A_0 A_i \mathrm{e}^{-\mathrm{i}(\Psi_0 - \Psi_i)}] \tag{3.8}$$

We may extract factor A_i from the square bracket. Apart from real constants, the result is

$$\mathrm{e}^{\mathrm{i}\Psi_i}[A_i + A_0 \mathrm{e}^{\mathrm{i}(\Psi_0 - \Psi_i)} + A_0 \mathrm{e}^{-\mathrm{i}(\Psi_0 - \Psi_i)}] \tag{3.9}$$

The first two terms here are identical with the field that exposed the plate; the first term corresponds to the coherent background and the second to the wave scattered by the object. An observer looking through the plate would therefore seem to see the object located in its original position. Except for its intensity, this reconstruction is theoretically identical to the object.

The third term is identical with the second term, apart from the sign of the phase term ($\psi_0 - \psi_i$). This term corresponds to a second reconstruction, located on the opposite side of the plate. This conjugate reconstruction is always present and is out of focus when we focus on the primary reconstruction. Its presence therefore degrades the primary reconstruction and is a major obstacle to the production of high-quality holograms. For practical purposes, this obstacle was removed when the laser provided sufficient coherence to separate the object wave from the coherent background with a prism or beam splitter.

3.1.2 Classification of Holograms

A hologram recorded on a photographic plate and processed normally is equivalent to a grating with a spatial varying transmittance. However, with suitable processing, it is possible to produce a spatial varying phase shift. In addition, if the thickness of the recording medium is large compared to the fringe spacing, volume effects are important. In an extreme case, it is even possible to produce holograms in which the fringes are planes running almost parallel to the surface of the recording material; such holograms can reconstruct an image in reflected light.

Based on these characteristics, holograms recorded in a thin recording medium can be divided into amplitude holograms and phase holograms. Holograms can be recorded in relatively thick recording media or can be classified either as transmission amplitude holograms, transmission phase holograms, reflection amplitude holograms, or reflection phase holograms.

3.1.3 Rainbow Holography

A multicolor image can be produced by a hologram recorded with three suitably chosen wavelengths. The resulting recording can be considered as made up of three incoherently superposed holograms. When it is illuminated once again with the three wavelengths used to make it, each of these wavelengths is diffracted by the hologram recorded with it to give a reconstructed image in the corresponding color. The superposition of these images yields a multicolor reconstruction.

The widely used lasers for rainbow holography are the He-Ne laser (λ=633nm) and the Ar^+ laser with two strong output lines (λ=514nm and 488nm). There are also other laser lines can be used which provide a better choice. One of these is the He-Cd laser line (λ=422nm), which is very attractive as a blue primary but involves the use of one more laser. The Kr^+ laser can be used for recording large holograms, because the power available is much higher, and single-mode output can be obtained with an etalon. The use of the Kr^+ laser line at 521 nm, or the output from a frequency-doubled Nd:YAG laser (λ=532nm), as the green primary has been found to give much better yellow images.

- **A typical method to generate the color rainbow hologram**

Fig.3.1 shows the way how to generate the color rainbow hologram in a single step. After those three light expose the recording holograms for three times respectively, the rainbow hologram can be formed in the dry plate. When this multiplexed hologram is illuminated with a light source, it reconstructs three superimposed images of the object.

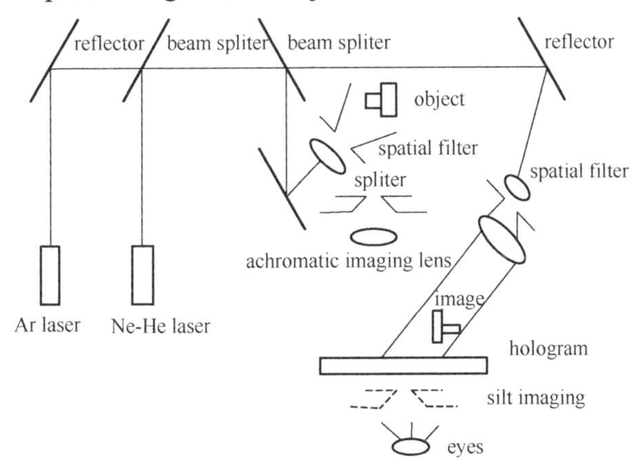

Fig.3.1 Color rainbow holography in a single step

- **The cross-talk problem**

A problem in multicolor holography is that each hologram diffracts, in addition to light of the wavelength used to record it, the other two wavelengths are as well. As a result, a total of nine primary images and nine conjugate images are produced. Three of these give rise to a full-color reconstructed image at the position originally occupied by the object. The remaining images resulting from light of one wavelength diffracted by a component hologram recorded with another wavelength are formed in other positions and overlap with and degrade the multicolored imagc.

Several methods have been tried to eliminate these cross-talk images, including spatial-frequency multiplexing, spatial multiplexing or coded reference waves, and division of the aperture field. However, all these methods suffer from drawbacks such as a restricted image field, a reduction in resolution, or a decrease in a signal-to-noise ratio (SNR). In addition, they need multiple laser wavelengths (or equivalent monochromatic light sources) to illuminate the hologram.

- **Volume holograms**

The first method to eliminate cross-talk that did not involve such penalties was based on the use of volume holograms. A hologram recorded with several wavelengths in a thick medium contains a set of regularly spaced fringe planes for each wavelength. When this hologram is illuminated once again with the original multiwavelength reference beam, each wavelength is diffracted with maximum efficiency by the set of fringe planes created originally by it, producing a multicolored image. However, the cross-talk images are severely attenuated since they do not satisfy the Bragg condition.

This principle was first applied to produce a two-color hologram of a transparency, and subsequently extended to three-color imaging of diffusely reflecting objects. The optical setup is shown schematically in Fig.3.2.

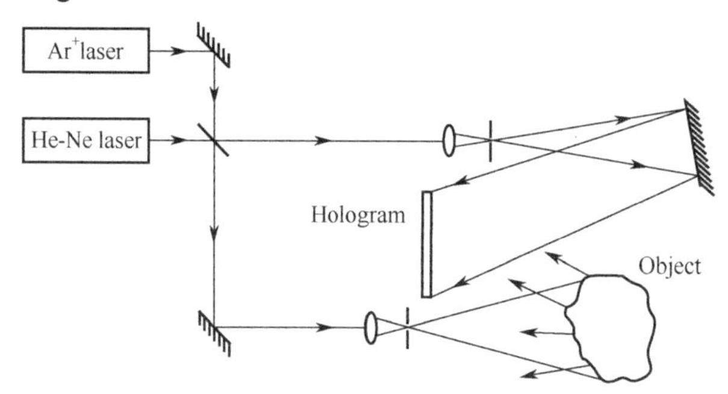

Fig.3.2 Setup used to record a multicolor hologram of a diffusely reflecting object in a thick recording medium

Blue and green light (λ=488nm and 514nm) from an Ar^+ laser was mixed with red light (λ=633nm) from a He-Ne laser to produce two beams containing light of all three wavelengths. One beam was used to illuminate the object while the other was used as a reference beam, and the resulting hologram was recorded in a thick photographic emulsion. When this hologram was illuminated once again with a similar multicolor beam at the appropriate angle, a multicolor reconstructed image with negligible cross-talk was obtained.

- **Multicolor rainbow holograms**

A completely different approach was opened up by the extension of the rainbow hologram technique to three-color recording. This made it possible to produce holograms that reconstruct very bright multicolor images when illuminated with a white-light source. Multicolor rainbow holograms can be produced in both one single step and two steps. Here a typical optical system for this purpose, using a concave mirror, is shown in Fig.3.3.

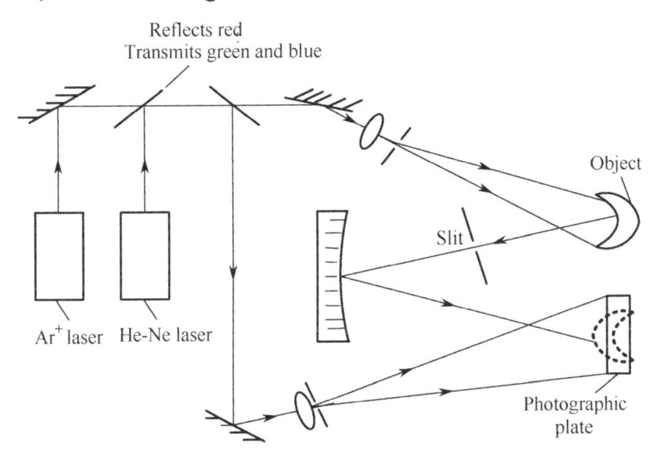

Fig.3.3 Layout of the optical system used to produce multicolor rainbow holograms in a single step with a concave mirror

The object, turned sideways, was placed on one side of the axis of the mirror so that its image was formed on the other side, at the same distance from the mirror. A vertical slit was placed between the object and the mirror, at such a distance from the mirror that a magnified image of the slit was formed in the viewing space at a distance of about 1m from the hologram.

Although in principle the three superimposed holograms making up the final multicolor rainbow hologram can be recorded on a single plate, there are several advantages in using a sandwich technique. Besides making it possible to use different types of photographic plates, whose characteristics are optimized for the different wavelengths, it also makes it much easier to match the diffraction efficiencies of the three individual holograms. In addition, it also gives much brighter images, since, with bleached holograms, the loss in diffraction efficiency due to multiplexing three holograms on a single plate can be partially avoided.

Multicolor holograms give bright images even when illuminated with an ordinary tungsten lamp. In addition, the images exhibit high color saturation and are free from cross-talk. Problem with emulsion shrinkage are eliminated, since volume effects are not involved. As with any rainbow hologram, the colors of the image change with the viewing angle in the vertical plane. This change can be utilized effectively in some types of displays, but, where necessary, it can be kept within acceptable limits by optimization of the length of the spectra projected into the viewing angles in the vertical plane.

3.1.4 Computer-generated Holography

Holograms generated by means of a computer or computer-generated holography (CGH), as shown in Fig.3.4, can be used to produce wavefronts with any prescribed amplitude and phase distribution; they are therefore extremely useful in applications such as laser-beam scanning and optical spatial filtering as well as for testing optical surfaces.

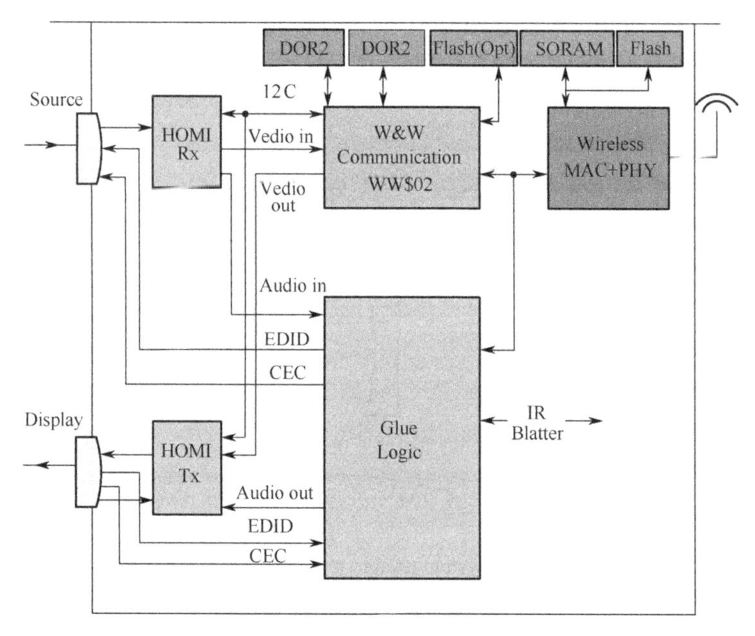

Fig.3.4 The principle of CGH

One of the most popular approaches to generate CGH involves two principal steps. The first step is to calculate the complex amplitude of the object wave at the hologram plane; for convenience, this is usually taken to be the Fourier transform of the complex amplitude in the object plane. It can be shown, by means of the sampling theorem, that if the object wave is sampled at a sufficiently large number of points, this can be done with no loss of information. Thus, if an image consisting of $N \times N$ resolvable elements is to be reconstructed, the object wave is sampled at $N \times N$ equally spaced points, and the $N \times N$ complex coefficients of its discrete Fourier transform are evaluated. This can be done quite easily with a computer program using the fast Fourier transform (FFT) algorithm for arrays containing as many as 1024×1024 points.

The second step involves using the computed values of discrete Fourier transform to produce a transparency (the hologram), which reconstructs the object wave when it is suitably illuminated. An alternative approach, which is possible only with a computer-generated hologram, is to produce a transparency that records both the amplitude and the phase of the object wave in the hologram plane. This can be thought of as the superimposition of two transparencies, one of constant thickness with a transmittance proportional to the amplitude of the object wave, and the other with thickness variations corresponding to the phase of the object wave, but no transmittance variations. Such a hologram has the advantage that it forms a single, on-axis image. In either case, the computer is used to control a plotter that produces a large scale version of the hologram. This master is photographically reduced to produce the required transparency.

- **Binary detour-phase hologram**

Although it is possible to use an output device with gray scale capabilities to produce the hologram, a considerable simplification results if the amplitude transmittance of the hologram has only two levels—either zero or one. Such a hologram is called a binary hologram.

The best known hologram of this type is the binary detour-phase hologram, which is made without explicit use of a reference wave or bias. To understand how this method of encoding the phase works, consider a rectangular opening ($a \times b$) in an opaque sheet (the hologram) centered on the origin of coordinates, as shown in Fig.3.5, which is illuminated with a uniform coherent beam of light of unit amplitude.

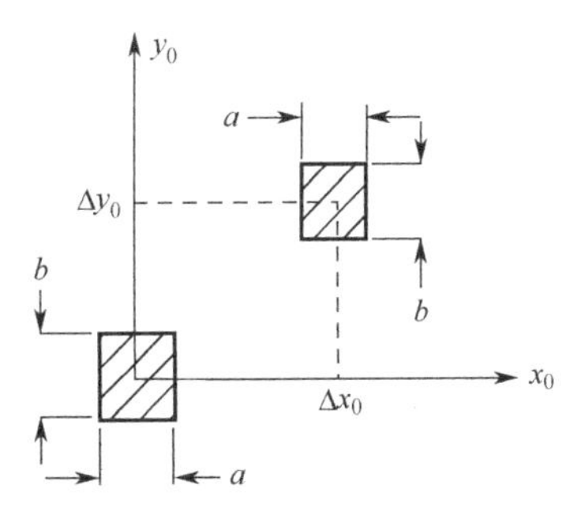

Fig.3.5 Diffraction at a rectangular aperture

The complex amplitude $U(x_i, y_i)$ at a point (x_i, y_i) in the diffraction pattern formed in the far field is given by the Fourier transform of the transmitted amplitude and is

$$U(x_i, y_i) = ab \sin \mathrm{c}(by_i / \lambda z) \tag{3.10}$$

We now assume that the centre of the rectangular opening is shifted to a point (Δx_0, Δy_0), and the sheet is illuminated by a plane wave incident at an angle. If the complex amplitude of the incident wave at the sheet is $\exp[i(\alpha\Delta x_0+\beta\Delta y_0)]$, the complex amplitude in the diffraction pattern becomes

$$\begin{aligned} U(x_i, y_i) &= ab \ \mathrm{sinc}\,(ax_i / \lambda z)\ \mathrm{sinc}\,(by_i / \lambda z) \exp\left[\mathrm{i}\left(\alpha + \frac{2\pi x_i}{\lambda z}\right)\Delta x_0 + \mathrm{i}\left(\beta + \frac{2\pi y_i}{\lambda z}\right)\Delta y_0\right] \\ &= ab\ \mathrm{sinc}\,(ax_i / \lambda z)\ \mathrm{sinc}\,(by_i / \lambda z)\exp[\mathrm{i}\alpha\Delta x_0 + \mathrm{i}\beta\Delta y_0]\exp\left[\mathrm{i}\left(\frac{2\pi x_i}{\lambda z}\right)\Delta x_0 + \frac{2\pi y_i}{\lambda z}\Delta y_0\right] \end{aligned} \tag{3.11}$$

If $ax_i \ll \lambda z$, $by_i \ll \lambda z$, Eq.(3.11) reduces to

$$U(x_i, y_i) = ab \exp[\mathrm{i}\alpha\Delta x_0 + \mathrm{i}\beta\Delta y_0]\exp\left[\mathrm{i}\left(\frac{2\pi x_i}{\lambda z}\right)\Delta x_0 + \frac{2\pi y_i}{\lambda z}\Delta y_0\right] \tag{3.12}$$

If, then, the computed complex amplitude of the object wave at a point in the hologram plane can be written as

$$o(n\Delta x_0, m\Delta y_0) = |o(n\Delta x_0, m\Delta y_0)| \exp[\mathrm{i}\phi(n\Delta x_0, m\Delta y_0)] \tag{3.13}$$

its modulus and phase at this point can be encoded, by making the area of the opening located in this cell equal to the modulus so that

$$ab = |o(n\Delta x_0, m\Delta y_0)| \tag{3.14}$$

and displacing the centre of the opening from the centre of the cell by an amount δx_{nm} given by the relation

$$\delta x_{nm} = (\Delta x_0 / 2\pi)\phi(n\Delta x_0, m\Delta y_0) \tag{3.15}$$

The total diffracted amplitude in the far field, which is obtained by summing the complex amplitudes due to all the $N \times N$ openings, Considering the dimensions of the cells and the angle of illumination are chosen ($\alpha\Delta x_0=2\pi$, $\alpha\Delta y_0=2\pi$, $\delta x_{nm} \ll \lambda z$), Eq.(3.12) can be written as

$$U(x_i, y_i) = \sum_{n=1}^{N}\sum_{m=1}^{N} |o(n\Delta x_0, m\Delta y_0)| \exp[\mathrm{i}\phi(n\Delta x_0, m\Delta y_0)] \exp\left[\mathrm{i}\frac{2\pi}{\lambda z}(nx_i\Delta x_0 + my_i\Delta y_0)\right] \tag{3.16}$$

which is the discrete Fourier transform of the computed complex amplitude in the hologram plane or, in other words, the described reconstructed image.

Binary detour-phase holograms have several attractive features. It is possible to use a simple pen- and -ink plotter to prepare the binary master, and problems of linearity do not arise in the photographic reduction process. Their chief disadvantage is that they are very wasteful of plotter resolution, since the number of addressable plotter points in each cell must be large to minimize the noise due to quantization of the modulus and the phase of the Fourier coefficicnts. When the number of phases-quantization lcvels is large, this noise is effectively spread over the whole image field, independent of the form of the signal. However, when the number of phase-quantization levels is small, the noise terms become shifted and self-convolved version of the signal, which are much more annoying.

- **Computer-generated interferograms**

Problems can arise with detour-phase holograms when encoding wavefronts with large phase variations, since a pair of apertures near the crossover may overlap when the phase of the wavefront moves through a multiple of 2π radians. This difficulty has been avoided in an alternative approach to the production of binary holograms based on the fact that an image hologram of a wavefront that has no amplitude variations is essentially similar to an interferogram, so that the exact locations of the transparent elements in the binary hologram can be determined by solving a grating equation.

Different methods can then be used to incorporate information on the amplitude variation in the object wavefront into the binary fringe pattern. In one method, the two-dimensional nature of the Fourier transform hologram is used to record the phase information along x direction, and the fringe heights in the y direction are adjusted to correspond to the amplitude. In another, the phase and the amplitude are recorded through the position and the width of the fringes along the direction of the carrier frequency, and in the third, the phase and amplitude of the object wave are encoded by the superimposition of two phase-only holograms.

- **Optical application of CGHs**

Now the CGHs have been widely used in daily life. One of the main applications of CGHs is interferometric tests of aspheric optical surfaces. Normally, such tests would require either an aspheric reference surface or an additional optical element, commonly referred to as a null lens, which converts the wavefront produced by the element under test into a spherical or plane wavefront.

An optical system using a Twyman-Green interferometer in conjunction with a CGH to test an aspheric mirror is shown in Fig.3.6. The hologram is a binary representation of the interferogram that would be obtained if the wavefront from an ideal aspheric surface were to interfere with a titled plane wavefront, and is placed in the plane in which the mirror under test is imaged. The superimposition of the actual interference fringes and the CGH produces a moiré pattern which maps the deviation of the actual wavefront from the ideal computed wavefront.

The contrast of the moiré pattern is improved by spatial filtering. This is done by reimaging the hologram through a small aperture placed in the focal plane of the reimaging lens. The position of this aperture is chosen so that it passes only the transmitted wavefront from the mirror under test and the diffracted wavefront produced by illuminating the hologram with the plane reference wavefront. These two wavefronts can be isolated if, in producing the computer-generated hologram, the slope of the aspheric wavefront along the same direction. Typical fringe patterns obtained with an aspheric surface, with and without a computer-generated hologram, are shown in Fig.3.7. Interference patterns obtained with an aspheric wavefront having a maximum slope of 35 waves per radius and a maximum departure of 19 waves from a reference sphere.

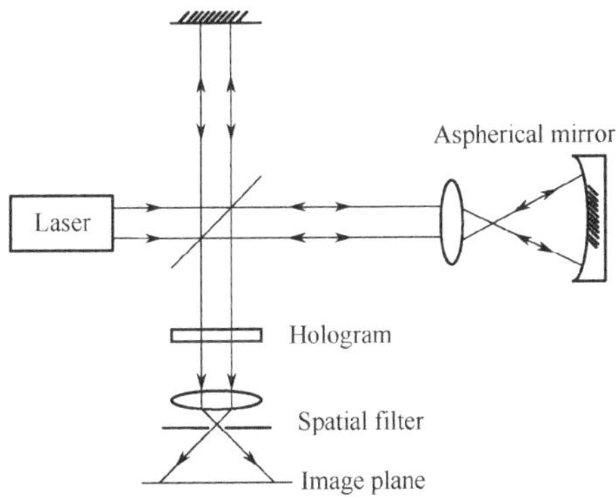

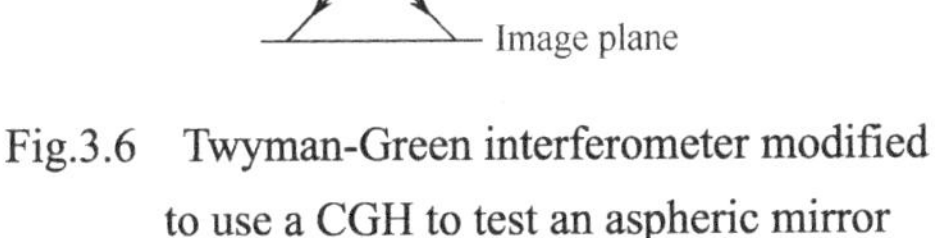

Fig.3.6 Twyman-Green interferometer modified to use a CGH to test an aspheric mirror

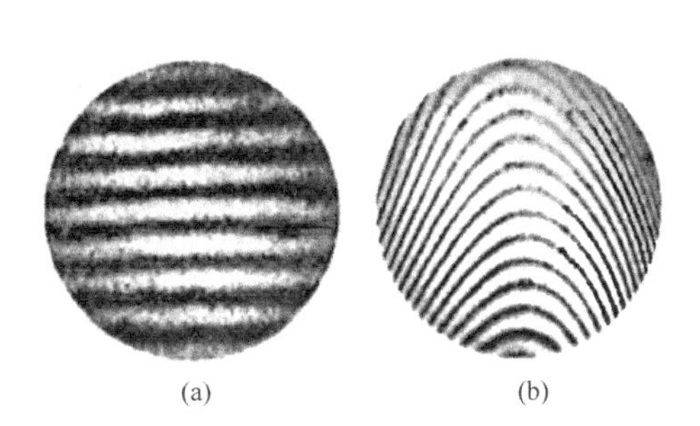

Fig.3.7 Comparison of interference patterns
(a) with (b) without a computer-generated hologram

Computer-generated holograms have been widely used to test aspheric surfaces. With the development of improved plotting routines and the application of techniques such as electron-beam recording and using layers of photoresist coated on optically worked substrates for the production of computer-generated holograms of very high quality, recent work has focused on refinements of the technique to obtain the highest precision. A preferred interferometer configuration is one in which both test and reference beams pass through the CGH so that aberrations of the substrate have no significant effect on the interferogram. It is also essential that the beams incident on the CGH should be collimated in order to reduce the effects of misalignment. Other factors to be considered are the use of image the test surface on the CGH and design of the CGH, using ray-tracing software, to compensate for off-axis aberrations introduced by the imaging system.

3.2 Wave-Optics Analysis of Optical Systems

3.2.1 Lens as a Phase Transformation

A lens is composed of an optically dense material, usually glass with a refractive index of approximately 1.5, in which the propagation velocity of an optical disturbance is less than the velocity in air. A lens is said to be a thin lens if a ray entering at coordinates (x, y) on one face exits at approximately the same coordinates on the opposite face, i.e. if there is negligible translation of a ray within the lens. Thus a thin lens simply delays an incident wavefront by an amount proportional to the thickness of the lens at each point.

Referring to Fig 3.8, let the maximum thickness of the lens (on its axis) be Δ_0, and let the thickness at coordinates (x, y) be $\Delta(x, y)$. Then the total phase delay suffered by the wave at coordinates (x, y) in passing through the lens may be written, i.e.,

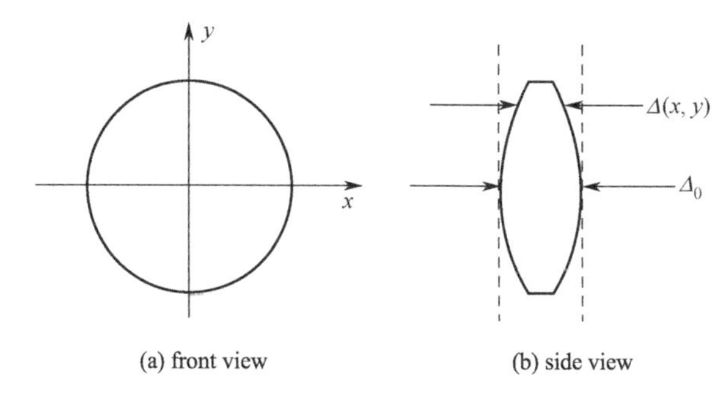

(a) front view (b) side view

Fig.3.8 The thickness function

$$\phi(x,y) = kn\Delta(x,y) + k[\Delta_0 - \Delta(x,y)] \tag{3.17}$$

where n is the refractive index of the lens material, $kn\Delta(x, y)$ is the phase delay introduced by the lens, and $k[\Delta_0 - \Delta(x, y)]$ is the phase delay introduced by the remaining region of free space between the two planes. Equivalently the lens may be represented by a multiplicative phase transformation of the form

$$t_l(x,y) = \exp[ik\Delta_0]\exp[ik(n-1)\Delta(x,y)] \tag{3.18}$$

The complex field $U_l'(x, y)$ across a plane immediately behind the lens is then related to the complex field $U_l(x, y)$ incident on a plane immediately in front of the lens by

$$U_l'(x,y) = t_l(x,y)U_l(x,y) \tag{3.19}$$

The problem remains to find the mathematical form of the thickness function $\Delta(x, y)$ in order that the effects of the lens may be understood.

- **The thickness function**

In order to specify the forms of the phase transformations introduced by a variety of different types of lenses, we first adopt a sign convention, i.e., as rays travel from left to right, each convex surface encountered is taken to have a positive radius of curvature, while each concave surface is taken to have a negative radius of curvature. Thus, in Fig. 3.9 (b) the radius of curvature of the left-hand surface of the lens is a positive number R_1, while the radius of curvature of the right-hand surface is a negative number R_2.

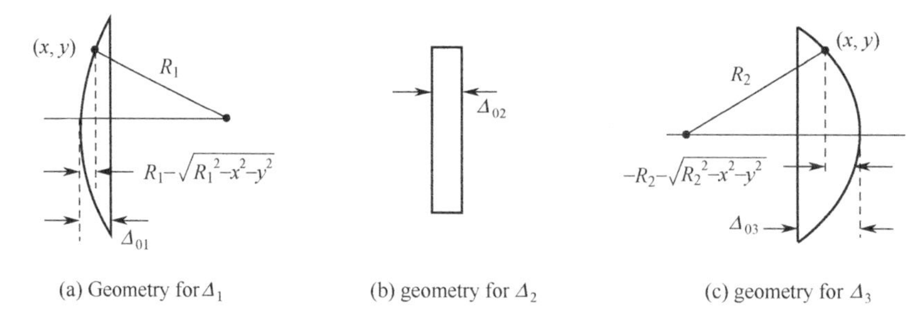

(a) Geometry for Δ_1 (b) geometry for Δ_2 (c) geometry for Δ_3

Fig.3.9 Calculation of the thickness function

To find the thickness $\Delta(x, y)$, we split the lens into three parts, as shown in Fig.3.9, and write the total thickness function as the sum of three individual thickness functions,

$$\Delta(x,y) = \Delta_1(x,y) + \Delta_2(x,y) + \Delta_3(x,y) \tag{3.20}$$

Referring to the geometries shown in that figure, the thickness function $\Delta(x, y)$ is given by

$$\Delta_1(x,y)=\Delta_{01}-(R_1-\sqrt{R_1^2-x^2-y^2})=\Delta_{01}-R_1\left(1-\sqrt{1-\frac{x^2+y^2}{R_1^2}}\right) \tag{3.21}$$

The second component of the thickness function comes from a region of glass of constant thickness Δ_{02}. The third component is given by

$$\Delta_3(x,y)=\Delta_{03}-(-R_2-\sqrt{R_2^2-x^2-y^2})=\Delta_{03}+R_2\left(1-\sqrt{1-\frac{x^2+y^2}{R_2^2}}\right) \tag{3.22}$$

where we have factored the positive number$-R_2$ out of the square root. Combining the three expressions for thickness, the total thickness is seen to be

$$\Delta(x,y)=\Delta_0-R_1\left(1-\sqrt{1-\frac{x^2+y^2}{R_1^2}})+R_2(1-\sqrt{1-\frac{x^2+y^2}{R_2^2}}\right) \tag{3.23}$$

where $\Delta_0=\Delta_{01}+\Delta_{02}+\Delta_{03}$.

- **The paraxial approximation**

The expression for the thickness function can be substantially simplified if attention is restricted to portions of the wavefront that lie near the lens axis, or equivalently, if only paraxial rays are considered. Thus we consider only values of x and y sufficiently small to allow the following approximations to be accurate

$$\sqrt{1-\frac{x^2+y^2}{R_1^2}}\approx 1-\frac{x^2+y^2}{2R_1^2}, \qquad \sqrt{1-\frac{x^2+y^2}{R_2^2}}\approx 1-\frac{x^2+y^2}{2R_2^2} \tag{3.24}$$

The resulting phase transformation will, of course, represent the lens accurately over only a limited area, but this limitation is no more restrictive than the usual paraxial approximation of geometrical optics. Note that the relations (3.24) amount to approximations of the spherical surfaces of the lens by parabolic surfaces. With the help of these approximations, the thickness function becomes

$$\Delta(x,y)=\Delta_0-\frac{x^2+y^2}{2}\left(\frac{1}{R_1}-\frac{1}{R_2}\right) \tag{3.25}$$

- **The phase transformation and its physical meaning**

Substitution of Eq.(3.25) into Eq.(3.18) yields the following approximation to the lens transformation

$$t_1(x,y)=\exp[\mathrm{i}kn\Delta_0]\exp\left[-\mathrm{i}k(n-1)\frac{x^2+y^2}{2}\left(\frac{1}{R_1}-\frac{1}{R_2}\right)\right] \tag{3.26}$$

The physical properties of the lens (that is, n, R_1 and R_2) can be combined in a single number f called the focal length, which is defined by

$$\frac{1}{f}=(n-1)\left(\frac{1}{R_1}-\frac{1}{R_2}\right) \tag{3.27}$$

Neglecting the constant phase factor, which we shall drop hereafter, the phase transformation may now be rewritten

$$t_1(x,y)=\exp\left[-\mathrm{i}\frac{k}{2f}(x^2+y^2)\right] \tag{3.28}$$

This equation will serve as our basic representation of the effects of a thin lens on an incident disturbance, where we neglect the finite extent of the lens.

The physical meaning of the lens transformation can best be understood by considering the effect of the lens on a normally incident, unit-amplitude plane wave. The field distribution U_l in front of the lens is unity, and Eqs.(3.19) and (3.28) yield the following expression for U_l' behind the lens

$$U_l'(x,y)=\exp\left[-\mathrm{i}\frac{k}{2f}(x^2+y^2)\right] \tag{3.29}$$

3.2.2 Frequency Analysis of Optical Imaging Systems

Considering the long and rich history of optics, the tools of frequency analysis and linear systems theory have played important roles for only a relatively short period of time. Nevertheless, in this short time these tools have been so widely and successfully used that they now occupy a fundamental place in the theory of imaging systems.

- **Generalized treatment of imaging systems**

Suppose that an imaging system of interest is composed, not of a single thin lens, but perhaps of several lenses, some positive, some negative, with various distances between them. The lenses need not be thin in the sense defined earlier. We shall assume, however, that the system ultimately produces a real image in space; this is not a serious restriction, for if the system produces a virtual image, to view that image it must be converted to a real image, perhaps by the lens of the eye.

To specify the properties of the lens system, we adopt the point of view that all imaging elements may be lumped into a single "black box", and that the significant properties of the system can be completely described by specifying only the terminal properties of the aggregate. Referring to Fig.3.10, the "terminals" of this black box consist of the planes containing the entrance and exit pupils. It is assumed that the passage of light between the entrance pupil and the exit pupil is adequately described by geometrical optics.

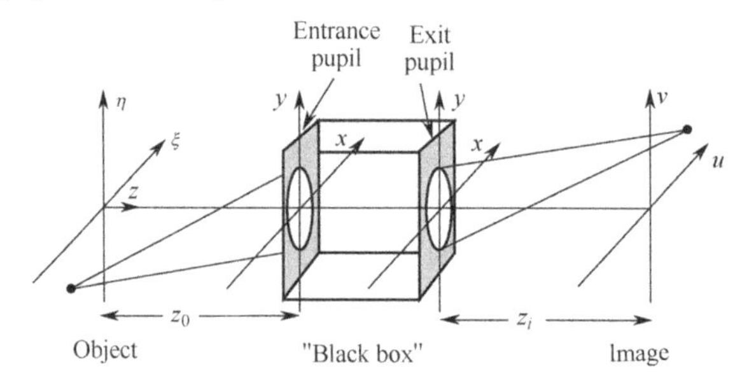

Fig.3.10 Generalized model of an imaging system

The entrance and exit pupils are in fact images of the same limiting aperture within the system. As a consequence, there are several different ways to visualize the origin of the spatial limitation of the wavefront that ultimately give rise to diffraction. It can be viewed as being caused by the physical limiting aperture internal to the system (which is the true physical source of the limitation). Equivalently it can be viewed as arising from the entrance pupil or from the exit pupil of the system.

We shall use the symbol z_0 to represent the distance of the plane of the entrance pupil from the object plane, and the symbol z_i to represent the distance of the plane of the exit pupil from the image plane. The distance z is then the distance that will appear in the diffraction equations that represent the effect of diffraction by the exit pupil on the point-spread function of the optical system. We shall refer either to the exit pupil or simply to the "pupil" of the system when discussing these effects.

An imaging system is said to be diffraction-limited if a diverging spherical wave, emanating from a point-source object, is converted by the system into a new wave, again perfectly spherical, that converges towards an ideal point in the image plane, where the location of that ideal image point is related to the location of the original object point through a simple scaling factor (the magnification), a factor that must be the same for all points in the image field of interest if the system is to be ideal. Thus the terminal property of a diffraction-limited imaging system is that a diverging spherical wave incident on the entrance pupil is converted by the system into a converging spherical wave at the exit pupil. For any real imaging system, this property will be satisfied, at best, over only finite regions of the object and image planes. If the object of interest is confined to the region for which this property holds, then the system may be regarded as being diffraction-limited.

Since geometrical optics adequately describes the passage of light between the entrance and exit pupils of a system, diffraction effects play a role only during passage of light from the object to the entrance pupil, or alternatively and equivalently, from the exit pupil to the image. It is, in fact, possible to associate all diffraction limitations with either of these two pupils. The two points of view that regard image resolution as being limited by: (1) the finite entrance pupil seen from the object space or (2) the finite exit pupil seen from the image space are entirely equivalent, due to the fact that these two pupils are images of each other.

- **Frequency response for diffraction-limited coherent imaging**

A coherent imaging system is linear in complex amplitude. This implies, of course, that such a system provides a highly nonlinear intensity mapping. If frequency analysis is to be applied in its usual form, it must be applied to the linear amplitude mapping. We would anticipate the transfer-function concepts applied directly to this system, provided it is done on an amplitude basis. To do so, define the following frequency spectra of the input and output, respectively

$$G_g(f_x, f_y) = \iint_{-\infty}^{+\infty} U_g(u, v) \exp[-i2\pi(f_x u + f_y v)] du dv \tag{3.30}$$

$$G_i(f_x, f_y) = \iint_{-\infty}^{+\infty} U_i(u, v) \exp[-i2\pi(f_x u + f_y v)] du dv \tag{3.31}$$

In addition, define the amplitude transfer function H as the Fourier transform of the space-invariant amplitude impulse response,

$$H(f_x, f_y) = \iint h(u, v) \exp[-i2\pi(f_x u + f_y v)] du dv \tag{3.32}$$

hence

$$G_i(f_x, f_y) = H(f_x, f_y) G_g(f_x, f_y) \tag{3.33}$$

Thus the effects of the diffraction-limited imaging system have been expressed, at least formally, in the frequency domain. It now remains to relate H more directly to the physical characteristics of the imaging system itself.

Finally we give some intuitive explanation as to why the scaled pupil function plays the role of the amplitude transfer function. Remember that in order to completely remove the quadratic phase factor across the object; the object should be illuminated with a spherical wave, in this case converging towards the point where the entrance pupil is pierced by the optical axis. The converging spherical illumination causes the Fourier components of the object amplitude transmittance to appear in thc entrance pupil, as well as in the exit pupil, since the latter is the image of the former. Thus the pupil sharply limits the range of Fourier components passed by the system.

- **Frequency response for diffraction-limited incoherent imaging**

In the coherent case, the relation between the pupil and the amplitude transfer function has been seen to be a very direct and simple one. When the object illumination is incoherent, the transfer function of the imaging system will be seen to be determined by the pupil again, but in a less direct and somewhat more interesting way. The theory of imaging with incoherent light has, therefore, a certain extra richness not present in the coherent case. We turn now to considering this theory; again attention will be centered on diffraction-limited systems, although the discussion that immediately follows applies to all incoherent systems, regardless of their aberrations.

Imaging systems that use incoherent illumination have been seen to obey the intensity convolution integral

$$I_i(u,v)=\kappa\iint|h(u-\tilde{\xi},v-\tilde{\eta})|^2 I_g(\tilde{\xi},\tilde{\eta})\mathrm{d}\tilde{\xi}\,\mathrm{d}\tilde{\eta} \tag{3.34}$$

Such systems should therefore be frequency-analyzed as linear mappings of intensity distributions. To this end, let the normalized frequency spectra of I_g and I_i be defined by

$$G_g(f_x,f_y)=\frac{\iint I_g(u,v)\exp[-\mathrm{i}2\pi(f_x u+f_y v)\mathrm{d}u\mathrm{d}v]}{\iint I_g(u,v)\mathrm{d}u\mathrm{d}v} \tag{3.35}$$

$$G_i(f_x,f_y)=\frac{\iint I_i(u,v)\exp[-\mathrm{i}2\pi(f_x u+f_y v)\mathrm{d}u\mathrm{d}v]}{\iint I_i(u,v)\mathrm{d}u\mathrm{d}v} \tag{3.36}$$

The normalization of the spectra by their "zero-frequency" values is partly for mathematical convenience, and partly for a more fundamental reason. It can be shown that any real and nonnegative function, such as I_g or I_i, has a Fourier transform which achieves its maximum value at the origin. We choose that maximum value as a normalization constant in defining G_g and G_i. Since intensities are nonnegative quantities, they always have a spectrum that is nonzero at the origin. The visual quality of an image depends strongly on the "contrast" of the image, or the relative strengths of the information-bearing portions of the image and the ever-present background. Hence the spectra are normalized by that background.

In a similar fashion, the normalized transfer function of the system can be defined by

$$H(f_x,f_y)=\frac{\iint|h(u,v)|^2\exp[-\mathrm{i}2\pi(f_x u+f_y v)]\mathrm{d}u\mathrm{d}v}{\iint|h(u,v)|^2\,\mathrm{d}u\mathrm{d}v} \tag{3.37}$$

Application of the convolution theorem to Eq.(3.34) then yields the frequency-domain relation

$$G_i(f_x,f_y)=H(f_x,f_y)G_g(f_x,f_y) \tag{3.38}$$

- **Comparison of coherent and incoherent imaging**

As seen in previous sections, the optical transfer function (OTF) of a diffraction-limited system extends to a frequency that is twice the cutoff frequency of the amplitude transfer function. It is tempting, therefore, to conclude that incoherent illumination will invariably yield "better" resolution than coherent illumination, given that the same imaging system is used in both cases. As we shall now see, this conclusion is in general not a valid one; a comparison of the two types of illumination is far more complex than such a superficial examination would suggest.

A major flaw in the above argument lies in the direct comparison of the cutoff frequencies in the two cases. Actually, the two are not directly comparable, since the cutoff of the amplitude transfer function determines the maximum frequency component of the image amplitude while the cutoff of the optical transfer function determines the maximum frequency component of image intensity. Surely any direct comparison of the two systems must be in terms of the same observable quantity, image intensity.

In the absence of a meaningful quality criterion, we can only examine certain limited aspects of the two types of images, realizing that the comparisons so made will probably bear little direct relation to overall image quality. Nonetheless, such comparisons are highly instructive, for they point out certain fundamental differences between the two types of illumination.

3.3 Optical Processing

Optical processing and related areas are important branches of modern optics and cannot easily be done justice in a few pages. Here we offer an introduction and concentrate as usual on a physical understanding of what can sometimes require enormously complicated mathematical treatments.

We begin by considering the standard optical processor of Fig.3.11. The object is illuminated by a coherent plane wave. Two identical lenses are placed in the locations shown in the figure. A simple, paraxial ray trace will show that the lenses project an inverted image into the focal plane of the second lens. Further, the principal planes lying between the two lenses are equidistant from object and image. The magnification of the processor is therefore 1.

Thus, the processor projects a real, inverted image at unit magnification.

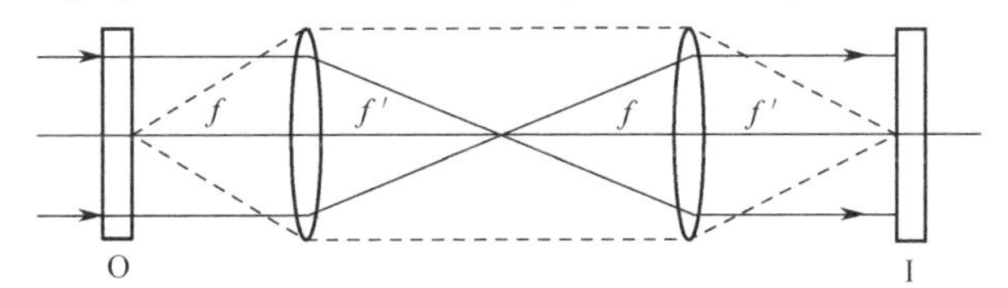

Fig.3.11 Coherent optical processor. Solid lines show the path of the undiffracted light; dashed lines show how the image is projected by the lenses

3.3.1 Abbe Theory

We now approach the processor from the point view of wave optics. The treatment that follows is identical to parts of the Abbe theory of microscopy, and we use the term, optical processing, to include many of the techniques of microscopy.

To simplify the explanation, the object is assumed to be a grating whose period is d, located in the input plane. The focal plane of the first lens is known as the frequency plane for reasons that will become clear later. We can find the distribution of intensity in the frequency plane by applying the grating equation

$$m\lambda = d\sin\theta, \quad m-0, \pm1, \pm2, \cdots \tag{3.39}$$

In paraxial approximation, $\sin\theta = \theta$, and principal maxima are located in the frequency plane at positions

$$\xi = m\lambda f' / d \tag{3.40}$$

where $\xi = 0$ corresponds to the intersection of the optical axis with frequency plane.

Consider, for the moment, any two adjacent diffraction orders, such as m=0,+1; assume that all other orders have somehow been suppressed. The second lens is then presented with two point sources a distance ξ apart. The intensity distribution in the output plane is, apart from a multiplicative constant

$$I' = \cos^2[(\pi/\lambda)\xi\sin\theta'] \tag{3.41}$$

where the prime refers to the output (image) plane. In paraxial approximation

$$\theta' = x'/f' \tag{3.42}$$

so that

$$I'(x') = \cos^2\left(\frac{\pi}{\lambda}\frac{\xi x'}{f'}\right) \tag{3.43}$$

The spacing d' between adjacent maxima of the fringes is found from the relation

$$\frac{\pi}{\lambda}\frac{\xi d'}{f'} = \pi \tag{3.44}$$

The value of ξ is $\lambda f'/d$ (since m=1). Thus,

$$d' = d \tag{3.45}$$

The grating in the input plane appears in the output plane as a sinusoidal interference pattern with precisely the same periodicity as the original grating.

If we allow the other orders to pass through the frequency plane, the grating becomes sharper than the sinusoidal fringe pattern and therefore more faithfully reproduces the object grating. We shall shortly return to this point in more detail. For now, let it suffice to say that the grating in the input plane will be recorded in the image plane only if the aperture in the frequency plane is large enough to pass both the 0 and either the +1 or −1 diffraction orders. It will be recorded as a sine wave, rather than as a square wave, but an image with the correct periodicity will nevertheless be detectable in the output plane.

Until now, we have tacitly assumed that the object is an amplitude transmission grating, consisting of alternately clear and opaque strips. We may take a glimpse of the power of optical processing if we allow the object to be a phase grating—that is, one that is transparent, but whose optical thickness varies periodically with the dimension x of the input plane. Ordinarily, the object is nearly invisible, because it is wholly transparent. Nevertheless, it exhibits diffracted orders, and one way (though not the best way) to render it visible would be to locate a special screen or spatial filter in the frequency plane and permit only the orders +1 and 0 to pass through holes in the screen. Then, as before, the image would be easily visible as a sinusoidal fringe pattern with the proper spacing between maxima. Spatial filtering is thus able to transform objects that are substantially invisible into visible images. This is the principle of phase microscopy, which is treated in more detail subsequently.

3.3.2 Fourier-Transform Optics

The amplitude distribution in the frequency plane is the Fraunhofer diffraction pattern of the object in the input plane. If the object is complicated and is characterized by an amplitude transmittance function $g(x)$, we must generalize our earlier Fraunhofer diffraction integral to include a factor $g(x)$ in the integrand,

$$E(\theta)=\frac{\mathrm{i}A}{\lambda}\int_{-b/2}^{b/2} g(x)\mathrm{e}^{-\mathrm{i}k\sin\theta x}\mathrm{d}x \tag{3.46}$$

or, in terms of the dimension ξ in the frequency plane

$$E(\xi)=\int_{-b/2}^{b/2} g(x)\mathrm{e}^{-\mathrm{i}\frac{2\pi\xi x}{\lambda f'}}\mathrm{d}x \tag{3.47}$$

apart from the multiplicative constants. If we define a new variable f_x

$$f_x=\frac{\xi}{\lambda f'} \tag{3.48}$$

we have

$$E(f_x)=\int_{-b/2}^{b/2} g(x)\mathrm{e}^{-\mathrm{i}2\pi f_x x}\mathrm{d}x \tag{3.49}$$

where $g(x)$ is a mathematical function that describes the object in the input plane and $E(f_x)$ is a representation of the electric-field amplitude in the frequency plane.

$E(f_x)$ greatly resembles the Fourier transform of $g(x)$. To make the resemblance perfect, we have only to define $g(x)$ to be zero outside the range $-b/2 < x < b/2$ and extend the range of integration from $-\infty$ to $+\infty$. The amplitude distribution in the frequency plane is then proportional to the Fourier transform $G(f_x)$ of $g(x)$

$$G(f_x)=\int_{-\infty}^{+\infty} g(x)\mathrm{e}^{-\mathrm{i}2\pi f_x x}\mathrm{d}x \tag{3.50}$$

We know from geometric optics that the processor casts an inverted image into the output plane. Thus, we define the positive x' axis to have opposite direction from the positive x axis. When the x' axis is defined in this way, the amplitude distribution $g'(x')$ in the output plane will be identical to that in the input plane.

The theory of the Fourier transform shows that the inverse transform

$$g'(x')=\int_{-\infty}^{+\infty} G(f_x)\mathrm{e}^{\mathrm{i}2\pi x' f_x}\mathrm{d}f_x \tag{3.51}$$

is also equal to $g(x)$. We thus conclude that the second lens performs the inverse transform, provided only that the x' axis be defined, as above, to take into account the fact that the system projects an inverted image. Needless to say, the fact can also be derived by rigorous mathematics.

3.3.3 Spatial Filtering

This term is usually used to describe manipulation of an image with masks in the frequency plane. We have already encountered some examples in connection with the Abbe theory and the Fourier series.

The simplest kind of spatial filter is a pinhole located in the focal plane of a lens. It acts as a low-pass filter and is commonly used to improve the appearance of gas-laser beams.

A gas-laser beam is typically highly coherent. The presence of small imperfection in a microscope objective, for example, results in a certain amount of scattered light. In an incoherent optical system this is of minor importance. Unfortunately, the scattered light in a coherent system interferes with the unscattered light to produce unsightly ring patterns that greatly resemble Fresnel zone plates.

Fortunately, the rings have relatively high spatial frequencies, so that these frequencies can be blocked by focusing the beam through a hole that transmits nearly the entire beam. The hole should be a few times the diameter of the Airy disk, so that very little other than the scattered light is lost.

Another important spatial filter is the high-pass filter. This consists of a small opaque spot in the center of the frequency plane. The spot blocks the low-frequency components of the object's spatial-frequency spectrum and allows the high-frequency components to pass. High-pass filtering can be used to sharpen photographs or to aid in examining fine detail.

Spatial filtering may also be used to remove unwanted detail from a photograph or to identify a character or a defect in a photograph. For example, suppose for some reasons that we had taken a photograph of a video display. The picture is composed of approximately 500 discrete, horizontal lines. The spacing of the lines determines the highest spatial frequency in the photograph. If we place the photograph in the input plane of the processor, we will see very strong diffraction orders in the frequency plane. These orders correspond to the harmonics of the grating formed by the horizontal lines.

To eliminate the lines in the output plane, we carefully insert two knife edges in the frequency plane. We locate the knife edges so that they cut off the +1 and −1 diffraction orders of the grating but pass all lower spatial frequencies. The result is that the picture is passed virtually unchanged, but the lines are eliminated completely. The picture has not been blurred, and the finest details are still visible in the output plane. Only the lines are absent.

Words and Expressions

a glimpse of	瞥见
adjacent	邻近的
aggregate	集合
aspheric	非球面的
attenuate	衰减

binary	二元的
bleached	漂白，褪色
branch	分支
cast into	映射，投射
coefficients	系数
conjugate	共轭的
convention	法则，习俗
converge	汇聚，汇聚于一点
convolution integral	卷积积分
cross-talk problem	交叉串扰
curve	曲线
degrade	使降级，使变差
dense	稠密的，浓厚的
detour-phase	迂回相位
deviation	偏离，偏差
discrete	不连续的，分离的，离散的
disturbance	扰动，干扰
domain	领域，范围
drawback	缺陷
emanate	发出，发射
emulsion	感光剂
encode	编码
encounter	遭遇
etalon	标准具
equivalently	等价地，相当于
extent	程度，范围，长度
exposure	曝光
factor	分解
faithfully	精确地，如实地
fine detail	细节
for the moment	现在，目前
Fourier-transform Optics	傅里叶变换光学
generalize	推广
gratings	光栅
harmonics	谐波，谐频
hereafter	今后，从此以后
holography	全息照相术
identify	识别
imperfection	不完美，瑕疵
implicitly	隐含地

instructive	有益的，有指导性的
interferogram	干涉图
interferometric	干涉仪的
intuitive	直觉的
invariant	不变的，不变量
inverse transform	逆变换
lump	集中
magnification	放大率
manipulation	处理
match	匹配
microscope objective	显微物镜
misalignment	非准直，不同轴
moiré	摩尔纹
multiplicative	乘法的
obstacle	障碍
on-axis	轴上
only if	只要当……时
optical processing	光信息处理
other than	除……之外
parabolic	抛物线形
penalty	问题，缺陷
phase-contrast microscopy	相衬显微镜
photoresist	光刻胶
plotter	绘图仪
pupil function	光瞳函数
prescribe	规定，指令，吩咐
principal maxima	主极大
proportional	成比例的
provided only	只要
quantization	量化
refinement	精加工
render	将……变为
resemble	类似，像
rigorous	严格的，精确的
sandwich technique	夹层技术
saturation	饱和度
scatter	散射
sharpen	锐化
shrinkage	收缩
sinusoidal	正弦类的

slope	斜率
spatial filter	空间滤波器
spatial filtering	空间滤波
strip	条状
substrate	基底，基片
suffice to	足以……
superimposition	重叠，叠加
suppress	压缩，抑制，消除，删除
term	术语
sampling theorem	抽样定理
scattered light	散射光
transmittance	透过率
transparency	透明度，透光度
tungsten lamp	钨丝灯
ultimately	最后，最终
utilize	利用，使用
undisturbed	不受扰动的
unsightly	难看的，丑的
volume effects	体效应
wavefront reconstruction	波前重构

Grammar 专业英语翻译方法（三）：数字的译法

英语的数词分为基数词、序数词、小数、分数、百分数和倍数词，一般多用作定语。

1. 数字的换算

thousand（$K=10^3$） million（$M=10^6$） billion（$B=10^9$） trillion（$T=10^{12}$） quadrillion（$Q=10^{15}$）

2. 数目不定的数词

在英语中，“ten”“hundred”“thousand”“million” 通常不加“s”，如果加“s”则表示为原数的若干倍。

tens of …，decades of …	几十，数十
dozens of …	几打，几十
scores of …	几十
hundreds of …	几百
thousands of …	几千
tens of thousands of …	几万
hundreds of thousands of …	几十万

tens of millions of … 几千万
a hundred and one 无数的，许多的
millions of … 千千万万
ten to one 十之八九
the seventies 七十年代
the early 1980's 二十世纪八十年代初期
up to 25% ≤25%
over (above) 50 五十多，高于五十
under (below) 70 七十以下，不到七十
sixty grams or so 六十克左右
on fifty-fifty basis 对等地，平分，各半
order of magnitude 数量级

—The result of measurement indicated that actual error probabilities were of the order of 1%.
测量结果表明，实际误差概率约为 1%。

3．增加量的译法

英汉两种语言对增加数量的说法的含义是有差别的，英语讲的增加倍数是包括基数在内，汉语则表示净增倍数，两者恰好相差 1 倍。

increased n times 增加到 n 倍（增加了 $n-1$ 倍）
increased by n times 增加了 n 倍
increase n%，increase by n% 增加 n%
increase by a factor of n 增加了 $n-1$ 倍
as … again as，again as … as 净增 1 倍
as much（many）again as 比……多 1 倍，2 倍
as … double（treble，quadruple …） 增加到 2（3、4）倍

4．减少量的译法

reduce（decrease）by a factor of n 减少到 $1/n$，减少了 $1-1/n$

—The old type aeroplanes fly twice more slowly than that of the new one.
旧式飞机的飞行速度只及新式飞机的一半。

5．数词短语的其他译例

from 11 to 21 从 11 到 21（不包括 11 和 21）
every other day 每隔一天
of the order of 约为，左右

—Careful investigation indicates that actual error probabilities were of the order of 1%.
仔细研究的结果表明，实际误差概率约为百分之一。

one hundred percent 百分之百，完全

—Great care must be taken to keep the mechanical drawings from errors by one hundred percent.

必须十分小心，务使图纸完全无误。

Second to none　　　　首屈一指

—The accuracy of this electronic computer is said to be second to none in the world.

这台电子计算机的准确性据说是世界上首屈一指的。

ten to one　　　　十之八九，很可能

—Ten to one，he has forgotten it.

他很可能已经忘了。

a thousand and one　　　　一千零一，许许多多

—In making that high-precision machine，the workers overcome a thousand and one difficulties.

在制造那台高精度机器时，工人们克服了许许多多的困难。

every *n* day　　　　每隔 *n*–1 天

—You should receive an injection every second day.

你必须每隔一天打一针。

Part 4 Lasers

A laser consists of a fluorescing material placed in a suitable optical cavity that is generally composed of two mirrors facing each other. Light passing through the fluorescing substance may be amplified by a process known as stimulated emission. If the material is properly prepared, stimulated emission can exceed absorption of the light. When sufficient amplification takes place, the property of the emission changes completely, resulting in a powerful, highly directional beam.

4.1 Amplification of Light

Suppose we locate an amplifying rod (or tube of gas or liquid) between two mirrors. One mirror is partially transparent; both are aligned parallel to one another and perpendicular to the axis of the rod. The optical length of the cavity thus formed is d, the reflectance of the partially transmitting mirror is R, and the (intensity) gain of the rod is G.

Initially, the only light emitted is that arising from fluorescence or spontaneous emission. As we have noted, the fluorescent emission is not directional, but some of this light will travel along the cavity's axis. The following heuristic argument shows how amplified fluorescence brings about laser emission.

Consider a wave packet that is emitted along the axis by a single atom. The packet undergoes many reflections from the mirrors. After each round trip in the cavity, it is amplified by G^2 and diminished by R. If the net round-trip gain exceeds 1,

$$G^2 R > 1 \tag{4.1}$$

then the wave grows almost without limit. Only waves that travel parallel to the axis experience such continuous growth; the result is therefore a powerful, directional beam. The useful output of the laser is the fraction that escapes through the partial reflector or output mirror.

There is a second condition necessary for lasing. Consider again the wave packet emitted along the axis. Its coherence length is great compared with the optical length of the cavity. In a sense, the atom therefore emits the packet over a finite time. Because of multiple reflections, the packet returns many times to the atom before the emission is completed. If the packet returns out of phase with the wave that is still being emitted by the atom, it will interfere destructively with that wave and effectively terminate the emission. We can, if we wish, say that such a wave has been reabsorbed, but the effect is as if the wave never existed.

The only waves that exist are, therefore, those for which constructive interference occurs,

$$m\lambda = 2d \tag{4.2}$$

As a result of the large number of reflections, only wavelengths quite close to $2d/m$ exist. This is analogous to the sharpness of multiple-beam interference fringes. In general, there are many such wavelengths within the fluorescent line-width of the source, so the value of d is not at all critical.

It is convenient now to consider the cavity separately from the amplifying medium. We may later combine their properties to account for the properties of the emitted light. In the following sections, we shall treat the properties of the amplifier and the cavity in somewhat greater detail.

4.2 Optical Resonators

An optical resonator consists of two mirrors facing each other, as in the Fabry-Perot interferometer. Both mirrors have highly reflecting coatings. Fabry-Perot interferometers have narrow transmissions pass bands at discrete optical frequencies.

4.2.1 Longitudinal Modes

To begin, consider a plane wave that originates inside a Fabry-Perot resonator, with amplitude A_0. After a large number of reflections have occurred, the total field inside the cavity is

$$E = A_0(1 + r^2 \mathrm{e}^{-\mathrm{i}\phi} + r^4 \mathrm{e}^{-2\mathrm{i}\phi} + \cdots) \tag{4.3}$$

for the wave traveling to the right. r is the amplitude reflection coefficient of both mirrors, and $\phi(=2kd)$ is the phase change associated with one round trip. The wave is assumed to travel normal to the mirrors.

We calculate the sum of the geometric series and find that the intensity I inside the cavity is

$$I = \frac{I_0}{1 + F\sin^2\dfrac{\phi}{2}} \tag{4.4}$$

where $F=4R/(1-R)^2$ and R is the reflectance. This is precisely the expression for the transmittance of a Fabry-Perot interferometer. I has value I_0 when the familiar condition for constructive interference satisfies, i.e.,

$$q\lambda=2d \tag{4.5}$$

where q is an integer.

Eq.(4.5) corresponds to a spacing of the two mirrors of q half-wavelengths. Each value of q corresponds to a different longitudinal mode of the interferometer having a frequency of

$$\nu = q\frac{c}{2d} \tag{4.6}$$

The frequency separation between modes follows from

$$\Delta\nu = \frac{c}{2d}(q+1) - \frac{c}{2d}q = \frac{c}{2d} \tag{4.7}$$

The frequency interval $\Delta\nu$ is called the free spectral range of the interferometer, or in terms of wavelength

$$\Delta\lambda = \frac{\lambda^2}{2d} \tag{4.8}$$

The longitudinal modes form an equally spaced comb of resonant frequencies, as shown in Fig.4.1.

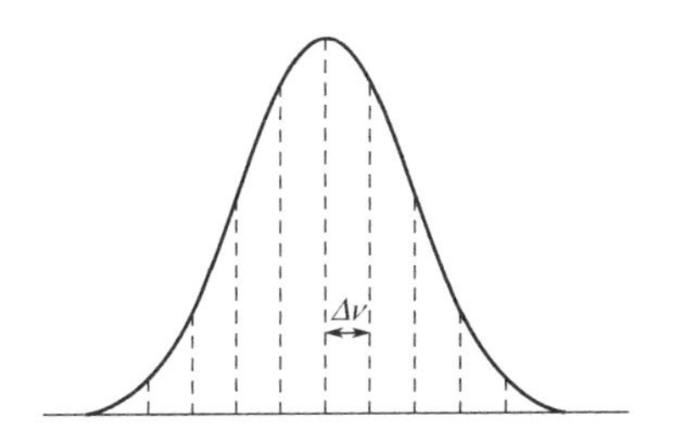

Fig.4.1 Longitudinal modes in a stable laser

4.2.2 Transverse Modes

These are most easily understood in terms of a cavity such as a confocal cavity. The confocal cavity has two identical mirrors with a common focus at F. Roughly speaking, we can define a transverse mode as the electric field distribution that is associated with any geometrical ray that follows a closed path (We will not consider cavities that do not allow a ray to follow a closed path; these are known as unstable resonators). Naturally, the ray will not describe the precise field distribution because of the effects of diffraction.

The simplest mode in a confocal cavity is described by a ray that travels back and forth along the axis. This is the 00 mode. Because of diffraction, the actual intensity distribution is that outlined. The output of a laser oscillating in this mode is a spherical wave with a Gaussian intensity distribution. The beam width is usually expressed as the radius w_0 at which the beam intensity falls to $1/e^2$ of its maximum value.

The next-simplest mode is also shown in Fig.4.2. This mode will oscillate only if the aperture (which is often defined by the laser tube) is large enough. Higher-order modes correspond to closed paths with yet-higher numbers of reflections required to complete a round trip. In Fig.4.2, Upper left corresponds to 00 mode; Upper right represents 01 mode; Lower left denotes 11 mode; Lower right corresponds to coherent superposition of two or more transverse modes.

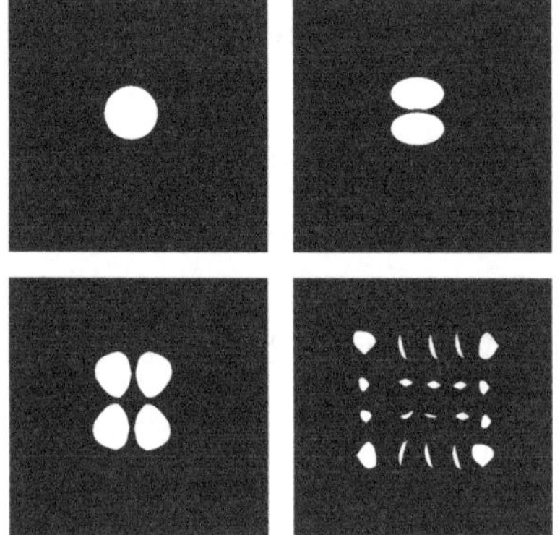

Fig.4.2 Laser transverse-mode patterns

Transverse-mode patterns are labeled according to the number of minima that are encountered when the beam is scanned horizontally (first number) and then vertically (second number). The modes shown all have rectangular symmetry; such rectangular modes characterize nearly all lasers, including those with cylindrical rods or tubes. In practice, the higher-order modes have greater loss (due to diffraction) than the 00 mode. If a laser oscillates in a certain high-order mode it also emits all modes with lower order. Such a multimode laser provides considerable power compared with that of a 00 or single-mode laser. Nevertheless, a single-mode laser is often desirable. In many gas lasers, the diameter of the laser tube is chosen small enough that diffraction loss prohibits oscillation of any mode other than the 00 mode.

4.2.3 Gaussian Beams

We have already noted that a laser oscillating in the 00 mode emits a beam with Gaussian intensity distribution. Higher-order modes also exhibit Gaussian intensity distributions, multiplied by certain polynomials (Hermite polynomials). Thus, beams emitted by any laser are called Gaussian beams.

In our earlier treatment of diffraction, we assumed that a uniform wavefront passed through the diffracting screen. We found, for example, that the far-field intensity distribution had an angular divergence of $1.22\lambda/D$ for a circular aperture. That result is not appropriate when Gaussian beams are used because then the intensity distribution is not uniform.

A detailed treatment of propagation of Gaussian beams is left to the references. Here we discuss the general results. The field distribution in any curved-mirror cavity is characterized by a beam waist; in a symmetrical cavity, the beam waist is located in the center of the cavity. The intensity distribution in the plane of the waist is Gaussian for the 00 mode; that is,

$$I(r) = \exp(-2r^2 / w_0^2) \tag{4.9}$$

where r is the distance from the center of the beam. For convenience, the intensity is normalized to 1 at the center of the beam. When $r = w_0$, the intensity is $1/e^2$ times the intensity at the center. Higher-order modes are characterized by the same Gaussian intensity distribution, but for the Hermite polynomials mentioned above. Fig.4.3 shows the transverse intensity distribution of a Gaussian beam for the 00 mode.

The beam propagates, both inside and outside the cavity, in such a way that it retains its Gaussian profile. That is, at a distance z from the beam waist, the intensity distribution is given by the preceding equation with w_0 replaced by $w(z)$, where

$$w(z) = w_0\left[1+\left(\frac{\lambda z}{\pi w_0^2}\right)^2\right]^{1/2} \tag{4.10}$$

At large distances z from the beam waist, the term in parentheses is large compared to 1. In this case, the Gaussian beam diverges with angle θ, where

$$\theta = \frac{\lambda}{\pi w_0} \tag{4.11}$$

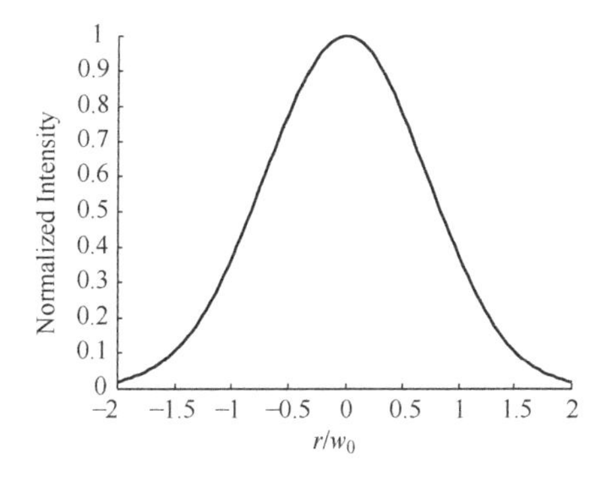

Fig.4.3 Transverse intensity distribution of a fundamental Gaussian beam

this is the far-field divergence of a Gaussian beam. If the beam is brought to a focus with a lens whose diameter is at least $2w_0$, the radius of the focal spot is $\lambda f' /(\pi w_0)$, which is somewhat smaller than the corresponding Airy-disk radius, $0.61\lambda f' /w_0$. In addition, the diffraction pattern is not an Airy disk, but has a Gaussian intensity distribution with no secondary maxima (unless the lens aperture vignettes a significant portion of the incident beam).

The radiation converges toward a beam waist and diverges away from it, as shown in Fig.4.4. Therefore, the wavefront must be planar at the beam waist. At a distance z from the waist, the radius of curvature $R(z)$ of the wavefront is

$$R(z) = z\left[1+\left(\frac{\pi w_0^2}{\lambda z}\right)\right] \tag{4.12}$$

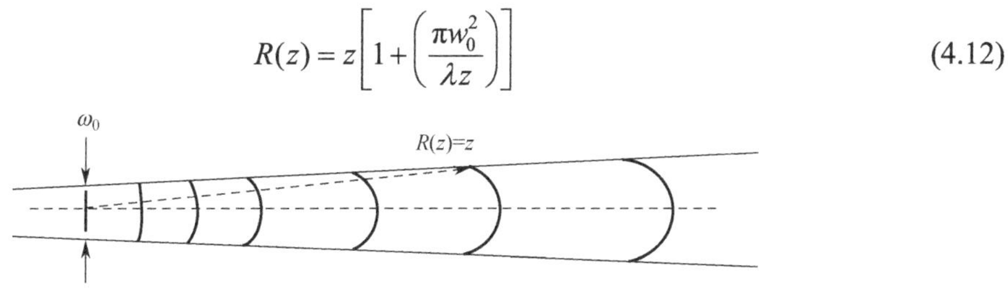

Fig.4.4 Contour of a Gaussian beam near $z = 0$

Only at great distance z from the waist does the beam acquire a radius of curvature equal to z.

Thus far, our comments have been general and apply to any Gaussian beam with a beam waist w_0 located at z=0. To apply the discussion to a laser cavity, we must know the size and position of the beam waist. In a confocal cavity whose mirrors are separated by d, w_0 is given by

$$w_0 = \sqrt{\frac{\lambda d}{2\pi}} \tag{4.13}$$

and the waist (z=0) is located in the center of the cavity, as with any symmetrical, concave-mirror cavity.

When the cavity is not confocal, it is customary to define stability parameters g_1 and g_2 by the equations

$$g_1 = 1 - d / R_1 \tag{4.14}$$

and

$$g_2 = 1 - d / R_2 \tag{4.15}$$

where R_1 and R_2 are the radii of curvature of the mirrors.

To find the size and location of the beam waist, we argue that the radius of curvature of the beam wavefront at the positions of the mirrors must be exactly equal to the radii of the mirrors themselves. If this were not so, then the cavity would not have a stable electric-field distribution; in terms of our earlier ray picture, the rays would not follow a closed path.

We therefore know the radius of curvature of the wavefront at two locations, which we may call z_1 (the distance from the waist to mirror 1) and z_2 (the distance from the waist to mirror 2). Setting $R(z_1)$=R_1 and $R(z_2)$=R_2, we may solve for z_1 and thereby locate the waist. The result is

$$z_1 = \frac{g_2(1-g_1)}{g_1 + g_2 - 2g_1 g_2} d \tag{4.16}$$

in terms of the stability parameters.

Similarly, we may solve for the beam waist w_0, which is, in general,

$$w_0 = \left(\frac{\lambda d}{\pi}\right)^{1/2} \left[\frac{g_1 g_2 (1 - g_1 g_2)}{g_1 + g_2 - 2g_1 g_2}\right]^{1/4} \tag{4.17}$$

We may now use the results of Gaussian-beam theory to find the radius $R(z)$ or the spot size $w(z)$ at any location z. In particular, at mirror 1,

$$w(z_1) = \left(\frac{\lambda d}{\pi}\right)^{1/2} \left[\frac{g_2}{g_1(1 - g_1 g_2)}\right]^{1/4} \tag{4.18}$$

and at mirror 2,

$$w(z_2) = \left(\frac{g_1}{g_2}\right)^{1/2} w(z_1) \tag{4.19}$$

Sometimes it is necessary to match the mode of one cavity to that of another; that is, the mode emitted by the first cavity must be focused on the second so that it becomes a mode of that cavity as well. For example, mode matching may be necessary when a spherical, confocal Fabry-Perot interferometer is used with a laser.

The simplest way to mode match two cavities a and b is to locate the point where their spot sizes $w(z)$ are equal. Then, calculate the beams' radii of curvature $R_a(z)$ and $R_b(z)$ at that point. A lens whose focal length f' is given by

$$1/f' = 1/R_a - 1/R_b \tag{4.20}$$

will match the radii of curvature of the two modes at the point.

Both radius and spot size must be matched to ensure effective mode matching. If both parameters are not matched, for example, power may be lost to higher-order modes in the second cavity. When it is not possible to match two cavities in the simple way described here, it may be necessary to expand or reduce the beam size with a lens before attempting mode matching with a second lens.

4.2.4 Resonator Configurations

The most commonly used laser resonators are composed of two spherical or flat mirrors facing each other, as shown in Fig.4.5. We will first consider the generation of the lowest order mode by such a resonant structure. Once the parameters of the TEM_{00} mode are known, all higher order modes simply scale from it in a known manner. Diffraction effects due to the finite size of the mirrors will be neglected in this section.

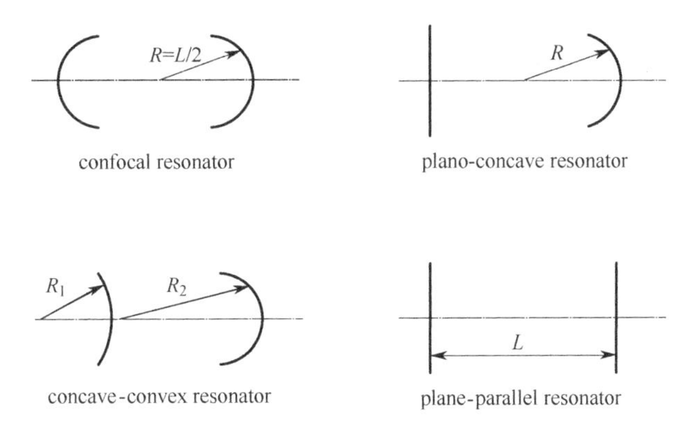

Fig.4.5 Common resonator configurations

Equations (4.17)~(4.19) treat the most general case of a resonator. There are many optical resonator configurations for which Eqs. (4.17)~(4.19) are greatly simplified.

- **Mirrors of equal curvature**

A special case of a symmetrical configuration is the concentric resonator that consists of two mirrors separated by twice their radius, that is, $R = L/2$. The corresponding beam consists of a mode whose dimension is fairly large at each mirror and focuses down to a diffraction-limited point at the center of the resonator. A concentric resonator is rather sensitive to misalignment, and the small spot can lead to optical damage.

Another very important special case of a resonator with mirrors of equal curvature is the confocal resonator. For this type of resonator the mirror separation equals the curvature of the identical mirrors, that is, $R = L$. The confocal configuration gives the smallest possible mode dimension for a resonator of given length. For this reason, confocal resonators are not often employed since they do not make efficient use of the active material.

- **Plano-concave resonator**

For a resonator with one flat mirror ($R_1=\infty$) and one curved mirror, it can be found that the beam waist w_0 occurs at the flat mirror.

A special case of this resonator configuration is the hemispherical resonator. The hemispherical resonator consists of one spherical mirror and one flat mirror placed approximately at the center of curvature of the sphere. The resultant mode has a relatively large diameter at the spherical mirror and focuses to a diffraction-limited point at the plane mirror. In practice, one makes the mirror separation d slightly less than R_2 so that a value of w_1 is obtained that gives reasonably small diffraction losses.

In solid-state lasers, the small spot size can lead to optical damage at the mirror. A near-hemispherical resonator has the best alignment stability of any configuration; therefore it is often employed in low-power lasers such as He-Ne lasers.

- **Concave-convex resonator**

A small radius convex mirror in conjunction with a large-radius concave or plane mirror is a very common resonator in high-average-power solid-state lasers. As follows from the discussion in the next section, as a passive resonator such a configuration is unstable.

However, in a resonator that contains a laser crystal, this configuration can be stable since the diverging properties of the convex mirror are counteracted by the focusing action of the laser rod. Since the convex mirror partially compensates for thermal lensing, a large mode volume can be achieved.

- **Plane-parallel resonator**

The plane-parallel or flat-flat resonator, which can be considered a special case of the large-radius mirror configuration if ($R_1= R_2= \infty$), is extremely sensitive to perturbation. However, in an active resonator, that is, a resonator containing a laser crystal, this configuration can be quite useful. Heat extraction leads to thermal lensing in the active medium; this internal lens has the effect of transforming the plane-parallel resonator to a curved mirror configuration. Therefore, the thermally induced lens in the laser material brings the flat-flat resonator into geometric stability.

4.2.5 Stability of Laser Resonators

The expressions for $w(z_1)$ and $w(z_2)$ contain a negative root of ($1-g_1g_2$). Unless the product g_1g_2 is less than 1, the spot sizes on the mirrors become infinite or imaginary. Laser cavities for which g_1g_2 exceeds 1 are unstable; those for which the product is just less than 1 are on the border of stability, because the spot size may exceed the mirror size and bring about great loss. Thus the stability criterion for lasers is

$$g_1g_2=\left(1-\frac{d}{R_1}\right)\left(1-\frac{d}{R_2}\right)<1 \tag{4.21}$$

The limiting case, $g_1g_2=1$, is a hyperbola. Stable resonators lie between the two branches of the hyperbola, unstable resonators, outside the two branches. Fig.4.6 shows the stability diagram for laser resonator. Resonators that are least sensitive to changes within the cavity (such as thermally induced focusing effects in solid or liquid lasers) lie on the hyperbola $g_1g_2=1/2$.

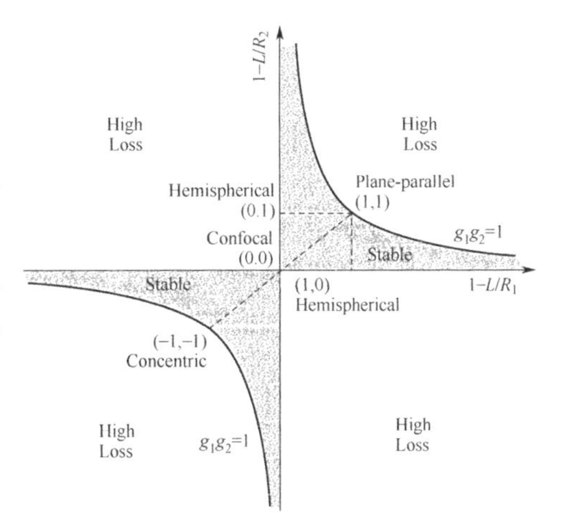

Fig.4.6 Stability diagram for laser resonator

4.3 Laser Amplifier

The power or energy from an oscillator with specific spatial, temporal, or spectral properties can be increased by adding one or more amplifying stages to the laser system. The main function of the amplifier is to increase the brightness of the beam. Amplification of pulsed or continuous wave (cw) power from an oscillator can be achieved using one of several techniques, including a master-oscillator power amplifier (MOPA) concept, a regenerative amplifier, a seeded or injection-locked power oscillator, and series connection of several gain modules within a common resonator. The choice of the specific amplifier approach depends on the power level, spectral and temporal characteristics of the input signal, as well as the desired output energy or power.

4.3.1 Pulse Amplification

The use of lasers as pulse amplifiers is of great interest in the design of high-energy, high-brightness radiation sources. In the pulse amplifiers, the input Q-switched or mode-locked pulse is considerably shorter than the fluorescent lifetime of the active medium. Hence the effect of spontaneous emission and pumping rate on the population inversion during the amplification process can be neglected. Furthermore, energy is extracted from the amplifier, which was stored in the amplifying medium, prior to the arrival of the pulse.

Consider the one-dimensional case of a beam of monochromatic radiation incident on the front surface of an amplifier rod of length l. The point at which the beam enters the gain medium is designated the reference point, $x = 0$.

If we ignore the effect of fluorescence and pumping during the pulse duration, we obtain for the population inversion

$$\partial n / \partial t = -\gamma n c \sigma \phi \tag{4.22}$$

where $\gamma = 1 + g_2/g_1$ (g_1 and g_2 are the degeneracy of energy levels) for a three level system and $\gamma = 1$ for a four-level system, c is the speed of light in the medium, σ denotes the emission cross section, and ϕ represents the photon density.

The growth of a pulse traversing a medium with an inverted population is described by the nonlinear, time-dependent photon-transport equation, which accounts for the effect of the radiation on the active medium and vice versa,

$$\frac{\partial \phi}{\partial t} = cn\sigma\phi - \frac{\partial \phi}{\partial x}c \tag{4.23}$$

The rate at which the photon density changes in a small volume of material is equal to the net difference between the generation of photons by the stimulated emission process and the flux of photons which flows out from that region.

Consider the one-dimensional case of a beam of monochromatic radiation incident on the front surface of an amplifier rod of length l. The point at which the beam enters the gain medium is designated the reference point, $x = 0$.

The two differential equations (4.22) and (4.23) must be solved for the inverted electron population n and the photon flux ϕ. Frantz and Nodvik solved these nonlinear equations for various types of input pulse shapes. The results are

$$E_{out} = E_s \ln\left\{1 + \left[\exp\left(\frac{E_{in}}{E_s}\right) - 1\right]\exp(g_0 l)\right\} \tag{4.24}$$

where E_{in} and E_{out} are input and output fluence (energy per unit area), respectively, and the saturation fluence E_s can be defined by

$$E_s = \frac{h\nu}{\gamma\sigma} = \frac{E_{st}}{\gamma g_0} \tag{4.25}$$

where E_{st} is the stored energy per volume, and g_0 is the small-signal gain coefficient.

The gain of the amplifier can be expressed as

$$G = \frac{E_s}{E_{in}} \ln\left\{1 + \left[\exp\left(\frac{E_{in}}{E_s}\right) - 1\right]G_0\right\} \tag{4.26}$$

Equation (4.26) permits one to calculate the gain of an amplifier as a function of the input energy density, provided that the small-signal gain or the energy stored in the amplifier is known.

In a laser system which has multiple stages, these equations can be applied successively, whereby the output of one stage becomes the input for the next stage.

4.3.2 Signal Distortion

As an optical signal propagates through a laser amplifier, distortions will arise as the result of a number of physical processes. We can distinguish between spatial and temporal distortions.

- **Spatial distortions**

The development of high-power laser oscillator-amplifier systems has led to considerable interest in the quality of the output beam attainable from these devices. Beam distortions produced during amplification may lead either to an increase in divergence or to localized high energy densities, which can cause laser rod damage. We will consider the main phenomena producing a spatial distortion, which is analogous to a wavefront or intensity distortion. The main factors producing the spatial distortions are as follows: (1) Nonuniform Pumping. Because of the exponential absorption of pump light, the center of the rod is pumped less than the edges, resulting in the spatial distortion of the signal. (2) Nonuniformities in the active material. Even laser rods with excellent optical quality contain a small amount of inherent stress, index of refraction variations, gradients in the active ion concentration, contaminants, inclusions, etc. These nonuniformities will significantly modify the energy distribution of an incoming signal. (3) Gain Saturation. A beam propagating through an amplifier can experience a distortion of the spatial profile because of the saturation-induced change in the distribution of gain. The weaker portions of the signal are amplified relatively more than the stronger portions because they saturate the medium to a lesser degree. (4) Diffraction Effects. Any limiting aperture in the amplifier section that removes energy at the edges of the beam will give rise to Fresnel rings. Diffraction effects are undesirable because they introduce strong intensity modulation in the beam that can produce spatial hot spots and lead to optical damage. (5) Thermal Distortions. The nonuniform pumping which leads to a higher gain coefficient at the edges of a laser rod also causes a nonuniform temperature profile across the rod. The absorbed pump power raises the temperature of the surface of the rod above the temperature of the rod center. As a result, a negative thermal lens is created, which distorts the wavefront of the beam.

- **Temporal distortions**

For a square pulse traversing an amplifier, the leading pulse edge sees a larger inverted population than does the trailing edge. This occurs simply because the leading edge stimulates the release of some of the stored energy and decreases the population inversion for the trailing edge. Thus, less energy is added to the final portions of a pulse than to the leading regions. The pulse-shape evolution depends appreciably on the rise time and shape of the leading edge of the input pulse. In general, one observes a forward shift of the peak as the pulse propagates through the amplifying medium.

Fig.4.7 illustrates the change of the pulse shape in an amplifier as a result of gain saturation. Curve 1 is the initially rectangular pulse, curves 2 and 3 are for g_0=0.1 cm^{-1}, l = 20 cm, E_{in}/E_s= 0.5, and E_{in}/E_s=0.1, respectively.

The intensity-induced index changes, are not only responsible for the catastrophic self-focusing problems in high-power laser amplifiers, but also cause a frequency shift of the pulse in the amplifying medium. In mode-locked pulses, the frequency modulation causes a broadening of the pulse width.

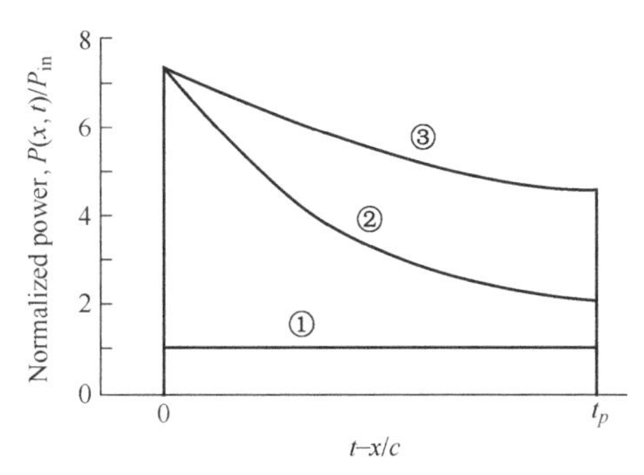

Fig.4.7 Change of the pulse shape in an amplifier as a result of gain saturation

4.3.3 Amplified Spontaneous Emission

The level of population inversion which can be achieved in an amplifier is usually limited by depopulation losses which are caused by amplified spontaneous emission (ASE). The favorable condition for strong ASE is a high gain combined with a long path length in the active material. In these situations, spontaneous decay occurring near one end of a laser rod can be amplified to a significant level before leaving the active material at the other end. A threshold for ASE does not exist; however, the power emitted as fluorescence increases rapidly with gain. Therefore, as the pump power increases, ASE becomes the dominant decay mechanism for the laser amplifier.

In high-gain, multistage amplifier systems, ASE may become large enough to deplete the upper state inversion. ASE is particularly important in large Nd:glass systems employed in fusion experiments. An analytical expression for the fluorescence flux I_{ASE} from a laser rod as a function of small signal gain, which has been found very useful in estimating the severity of ASE, is given in (4.27) with approximation $G_0 > 1$

$$\frac{I_{ASE}}{I_s} = \frac{\Omega}{4} \frac{G_0}{(\ln G_0)^{1/2}} \tag{4.27}$$

where I_s is the saturation flux, Ω is the solid angle around the axis of the active material and G_0 is the small signal gain of the active medium.

In Fig.4.8, the measured ASE from a four-stage, double-pass Nd:YAG amplifier chain is plotted as a function of diode-pump input, in which dots are measured values and solid line represents calculated values from Eq.(4.27). It can be seen that at an input of about 500mJ into each amplifier ASE starts to become noticeable, and quickly increases in intensity to reach 75 mJ at an input of 900 mJ per amplifier.

The detrimental effect of ASE can be seen in Fig.4.9, which shows the output from the amplifier chain as a function of pump input. As the amplified spontaneous emission increases, the slope of output versus input decreases and the difference can be accounted for by the loss due to ASE.

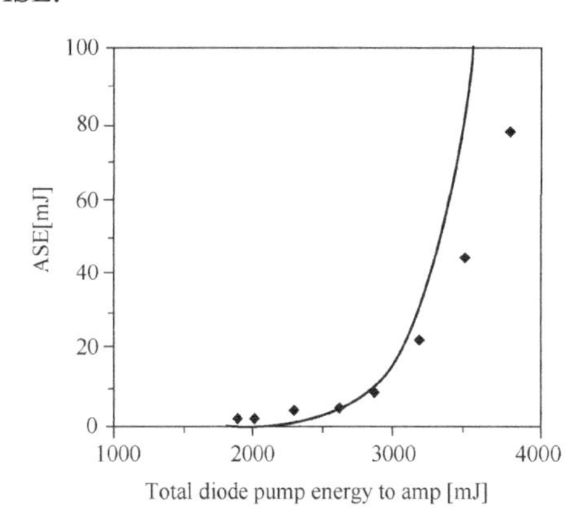

Fig. 4.8 ASE from a 4-stage double pass Nd: YAG amplifier chain

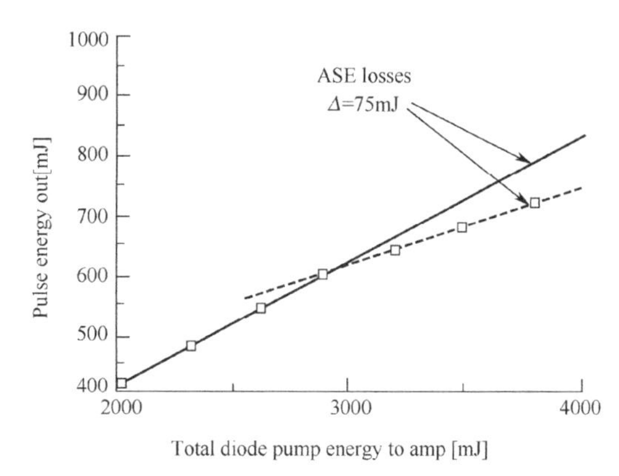

Fig. 4.9 Signal output versus pump input from a multistage Nd:YAG amplifier chain

So far, we discussed the amplification of fluorescence generated by the active material itself. It has been observed that flashlamp pump radiation, which is within the spectral region of the laser transition, will also be amplified and can lead to a reduction of stored energy in the laser material.

Gain saturation resulting from amplification of pump radiation in the active medium has been observed mainly in Nd:glass and Nd:YAG amplifiers. In addition to the depumping process caused by the lamp radiation, fluorescence being emitted from the laser rod escaping out into the laser pump cavity can be reflected back into the laser rod and thereby stimulating further off-axis emission. Thus the presence of pump-cavity walls, which have a high effective reflectivity at the fluorescence wavelength, causes a transverse depumping action which depletes the energy available for on-axis stimulated emission. Elimination of these effects may require the use of optical filters in the pump cavity, cladding of the laser rod with a material which absorbs at the laser wavelength, or the addition of chemicals to the cooling fluid which serve the same purpose.

4.4 Laser Techniques

4.4.1 Q-Switching

A mode of laser operation extensively employed for the generation of high pulse power is known as Q-Switching. Laser resonators are characterized by the quality factor Q, which is defined as the ratio of the energy stored in the cavity to the energy loss per cycle, i.e.,

$$Q = 2\pi\left[1 - \exp\left(-\frac{2\pi\omega_0}{\tau_c}\right)\right]^{-1} \tag{4.28}$$

where ω_0 denotes the angular frequency , τ_c represents all the losses in an optical resonator of a laser oscillator. Since τ_c has the dimension of time, the losses are expressed in terms of a relaxation time.

In the technique of Q-switching, energy is stored in the amplifying medium by optical pumping while the cavity Q is lowered to prevent the onset of laser emission. Although the energy stored and the gain in the active medium are high, the cavity losses are also high, lasing action is prohibited, and the population inversion reaches a level far above the threshold for normal lasing action. The time for which the energy may be stored is on the order of the lifetime of the upper level of the laser transition. When a high cavity Q is restored, the stored energy is suddenly released in the form of a very short pulse of light. Because of the high gain created by the stored energy in the active material, the excess excitation is discharged in an extremely short time. The peak power of the resulting pulse exceeds that obtainable from an ordinary long pulse by several orders of magnitude. The development of a Q-switched laser pulse is shown in Fig.4.10, in which the flashlamp output, resonator loss, population inversion, and photon flux as a function of time.

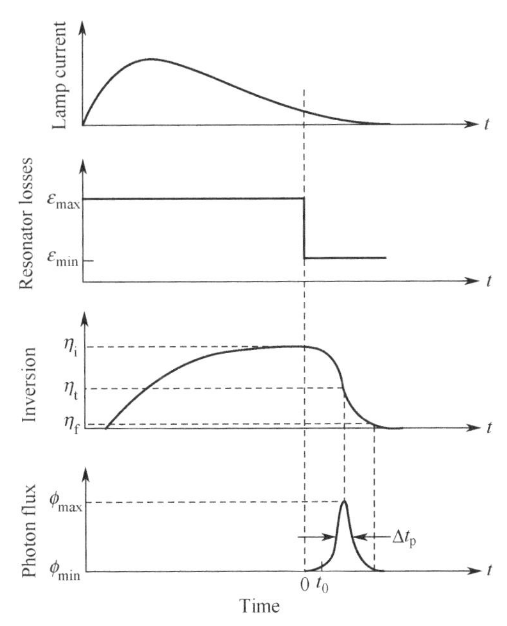

Fig.4.10 Development of a Q-switched laser pulse

Q-switches have been designed based upon

rotational, oscillatory, or translational motion of optical components. These techniques have in common that they inhibit laser action during the pump cycle by either blocking the light path, or causing a mirror misalignment, or reducing the reflectivity of one of the resonator mirrors. Near the end of the pumping pulse, when maximum energy has been stored in the laser medium, a high Q-condition is established and a Q-switch pulse is emitted from the laser.

4.4.2 Mode Locking

The nonlinear absorption of saturable absorbers was first successfully employed for simultaneously Q-switching and mode-locking solid-state lasers in 1965. The saturable absorbers consisted of organic dyes that absorb at the laser wavelength. At sufficient intense laser radiation, the ground state of the dye becomes depleted, which decreases the losses in the resonator for increasing pulse intensity.

In pulsed mode-locked solid-state lasers, pulse shortening down to the limit set by the gain-bandwidth is prevented because of the early saturation of the absorber which is a result of the simultaneously occurring Q-switching process. Shorter pulses and a much more reproducible performance are obtained if the transient behavior due to Q-switching is eliminated. In steady-state or cw mode locking, components or effects are utilized which exhibit a saturable absorber-like behavior, i.e., a loss that decreases as the laser intensity increases. The distinction between an organic dye suitable for simultaneous mode-locking and Q-switching, as opposed to only Q-switching the laser, is the recovery time of the absorber. If the relaxation time of the excited-state population of the dye is on the order of the cavity round trip, i.e., a few nanoseconds, passive Q-switching will occur. With a dye having a recovery time comparable to the duration of mode-locked pulses, i.e., a few picoseconds, simultaneous mode-locking and Q-switching can be achieved.

Fig.4.11 shows the output signal of an ideally mode-locked laser. The spectral intensities have a Gaussian distribution, while the spectral phases are identically zero. In the time domain the signal is a single Gaussian pulse. As can be seen from this figure, mode locking corresponds to correlating the spectral amplitudes and phases. When all the initial randomness has been removed, the correlation of the modes is complete and the radiation is localized in space in the form of a single pulse.

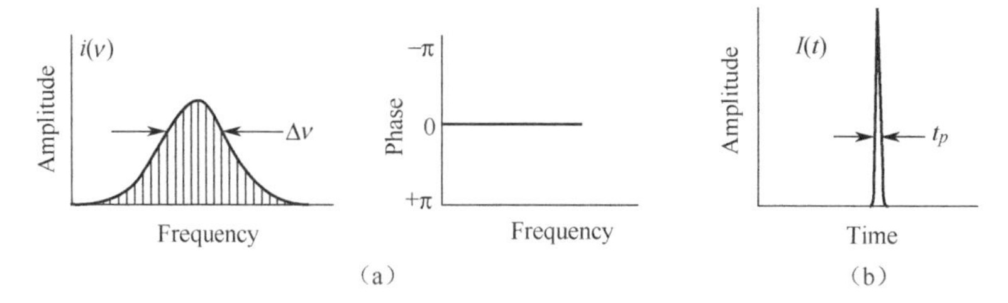

Fig.4.11 Signal structure of an ideally mode-locked laser,

(a) the spectral intensities have a Gaussian distribution, while the spectral phases are identically zero;

(b) in the time domain the signal is a transform-limited Gaussian pulse.

In recent years, several passive mode-locking techniques have been developed for solid-state lasers whereby fast saturable absorber-like action is achieved in solids. Most of these novel optical modulators utilize the nonresonant Kerr effect. The Kerr effect produces intensity-dependent changes of the refractive index. It is generally an undesirable effect because it can lead to self focusing and filament formation in intense beams. In contrast to the absorption in bleachable dyes, the nonresonant Kerr effect is extremely fast, wavelength-independent, and allows the generation of a continuous train of mode-locked pulse from a cw-pumped laser.

4.4.3 Mode Selecting

- **Transverse modes**

Many applications of lasers require operation of the laser at the TEM_{00} mode, since this mode produces the smallest beam divergence, the highest power density, and, hence, the highest brightness. Focusing a fundamental-mode beam by an optical system will produce a diffraction-limited shot of maximum power per unit area. Generally speaking, in many applications it is a high brightness rather than large total emitted power that is desired from the laser.

Furthermore, the radiation profile of the TEM_{00} mode is smooth. This property is particularly important at higher power levels, since multimode operation leads to the random occurrence of local maxima in intensity, so-called hot spots, which can exceed the damage threshold of the optical components in the resonator.

Transverse mode selection generally restricts the area of the laser cross section over which oscillation occurs, thus decreasing the total output power. However, mode selection reduces the beam divergence so that the overall effect of mode selection is an increase in the brightness of the laser.

Most lasers tend to oscillate not only in higher-order transverse modes, but in many such modes at once. Because of the fact that higher-order transverse modes have a larger spatial extent than the fundamental mode, a given size aperture will preferentially discriminate against higher-order modes in a laser resonator. As a result, the question of whether or not a laser will operate in the lowest-order mode depends on the size of this mode and the diameter of the smallest aperture in the resonator. If the aperture is much smaller than the TEM_{00} mode size, large diffraction losses will occur which will prevent the laser from oscillating. If the aperture is much larger than the TEM_{00} mode size, then the higher-order modes will have sufficiently small diffraction losses to be able to oscillate.

- **Longitudinal modes**

A typical laser will oscillate in a band of discrete frequencies which have an overall width of about 10^{-4} of the laser frequency. Although this is a rather monochromatic light source, there are still many applications for which greater spectral purity is required.

In one of the earliest attempts to narrow the spectral width of a laser, tilted Fabry-Perot etalons were employed as mode-selecting elements. Also the concept of axial mode selection based on an analysis of the modes of a multiple-surface resonator was introduced. Since then, many mode-selecting techniques have been developed, such as interferometric mode selection, enhancement of longitudinal mode selection, injection seeding etc..

—Interferometric mode selection

A Fabry-Perot-type reflector is inserted between the two mirrors of the optical resonator. This will cause a strong amplitude modulation of the closely spaced reflectivity peaks of the basic laser resonator and thereby prevent most modes from reaching threshold.

—Enhancement of longitudinal mode selection

In this case, an inherent mode-selection process in the resonator is further enhanced by changing certain system parameters, such as shortening the resonator, removal of spatial hole burning or lengthening of the Q-switch built-up time.

—Injection seeding

This technique takes advantage of the fact that stable, single longitudinal mode operation can readily be achieved in a very small crystal located within a short, traveling wave resonator. These devices which are end-pumped by a laser diode array are by themselves not powerful enough for most applications, unless the output of such a device is coupled into a large slave oscillator for amplification.

4.4.4 Frequency Control

Frequency control requires a certain amount of tunability of the laser emissions, either to lock the frequency to a reference or to an incoming signal for coherent detection, or to provide an output signal at a specific wavelength for spectroscopic studies.

In certain applications, such as coherent light detection and ranging (LIDAR) system, stabilized single-frequency laser source is required, both as local oscillator for the detection system, and as injection seeding source for a pulsed transmitter.

Miniature diode-pumped solid-state lasers, such as monolithic or microchip devices, are particularly attractive for this application because of their single-frequency operation. Furthermore, the short resonator of these devices supports a relatively large tuning range because longitudinal modes are widely separated. In monolithic lasers, the frequency can be tuned by thermal expansion or by mechanical stress applied to the crystal. Discrete-element miniature lasers can be tuned by insertion of an electro-optic modulator into the cavity or by cavity-length adjustment.

In monolithic lasers such as the microchip laser, or nonplanar ring laser, the resonator mirrors are coated directly on the laser crystal, which precludes the introduction of traditional intra-cavity elements. However, varying the temperature of the crystal will change the resonator length, and therefore frequency tunes the laser emission.

Instead of changing the temperature of the laser crystal by means of a TE cooler, pump power modulation has been successfully employed to cause rapid frequency tuning in miniature lasers. As the pump power increases, more thermal energy is deposited in the gain medium, raising its temperature and therefore changing the resonant frequency of the laser cavity. Modulation of the pump power has the undesirable effect of changing the laser output power. In implementing this technique, TE coolers maintain a fixed average temperature of the crystal. Around this temperature point, rapid thermal changes are induced by means of current modulation of the pump-laser diode. The thermal response is sufficiently fast, permitting phase locking of the laser.

Another technique for precisely tuning the laser frequency of a monolithic laser is based on stress-induced changes in the resonator. In these designs, stress is applied to the laser crystal by using a piezoelectric transducer. This results in a stress-induced birefringence of the refractive index, and a strain-related distortion or elongation of the resonator.

In the case of discrete-element microcavity lasers, frequency tuning has been achieved by cavity length modulation or by incorporating an electro-optic modulator into the resonator. In the former design, an air gap between the two pieces forming a ring resonator was adjusted by a piezoelectric transducer. Cavity length tuning was achieved over a frequency range of 13.5 GHz at a speed on the order of milliseconds.

Frequency stability is degraded by mechanical vibrations, acoustic noise, and ambient temperature variations. The most stringent requirement for absolute frequency controls is imposed on scientific lasers developed for gravitational-wave detectors. Absolute frequency stability can be achieved by locking the output of the laser to a high finesse reference Fabry-Perot cavity employing the Pound-Drever-Hall (PDH) servo system.

4.4.5 Wavelength Selection

In the operation of lasers with very broad gain curves, such as alexandrite, Ti:Sapphire, and Cr:GSGG, it is necessary to use a wavelength selection technique to (a) restrict laser action to a specified wavelength and (b) tune the laser output. Several different methods are available (in principle) for providing the wavelength selection and tuning. These include (a) use of a prism inside the resonator, (b) utilization of an adjustable optical grating within the laser, (c) use of intracavity etalons, or (d) use of one or more thin birefringent plates within the laser that are tilted at Brewster's angle.

The technique most commonly employed for the wavelength selection of tunable lasers is the birefringent filter. In its simplest form, the birefringent filter consists of a single thin birefringent crystal located inside the laser, as shown in Fig.4.12.

For simplicity, we assume that the birefringent axes lie in the plane of the crystal, and that the crystal is tilted at Brewster's angle. Wavelength selection occurs with the birefringent filter because of the two different crystal indices of refraction. When the laser light has a wavelength corresponding to an integral number of full-wave retardations, the laser operates as if the filter were not present. At any other wavelength, however, the laser mode polarization is modified by the filter and suffers losses at the Brewster surfaces.

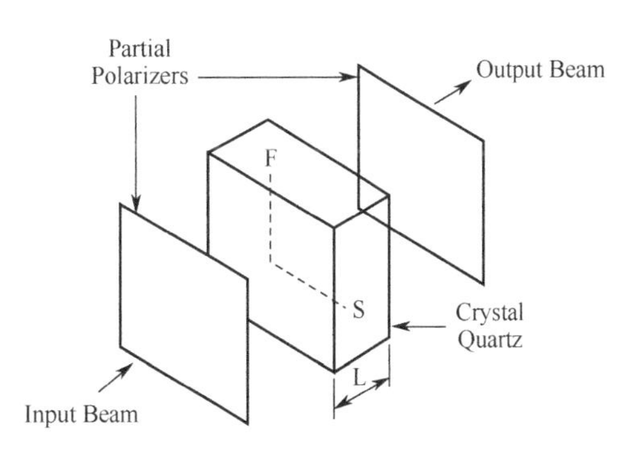

Fig.4.12 Single-crystal wavelength selector

Tunability of the laser is achieved by rotating the

birefringent crystal in its own plane. This changes the included angle φ between crystal optic axis and the laser axes and, hence, the effective principal refractive indices of the crystal. The amplitude transmittance of the single-stage filter of Fig.4.12 has been calculated. For a quartz crystal rotated to $\varphi = 45°$ and tilted to Brewster's angle, the transmittance at unwanted wavelengths is about 82%. This may or may not provide adequate suppression of unwanted wavelengths for certain lasers. One way to lower the filter transmittance in the rejection band is to use a stack of identical crystal plates that are similarly aligned, as shown in Fig.4.13. If one uses a stack of 10 quartz plates, the resulting transmittance in the rejection band is about 15%, which is certainly small enough to suppress unwanted laser frequencies.

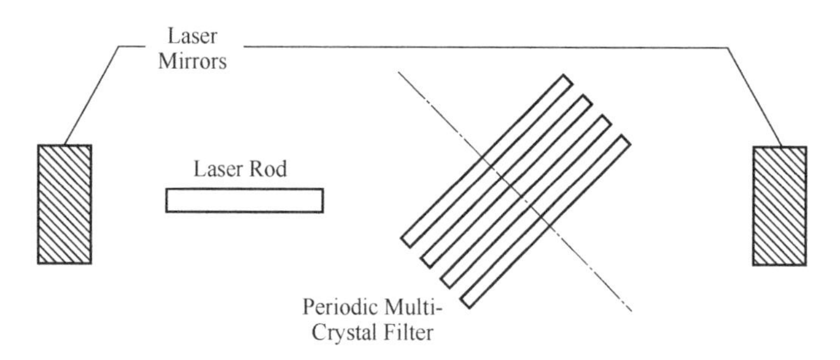

Fig.4.13 multiple-crystal wavelength selector

Another technique for lowering the filter transmittance in the rejection band is to include more Brewster's angle surfaces in the laser. Still another approach for narrowing the width of the central passband of the filter is to use several crystals in series whose thicknesses vary by integer ratios. The central passband of this kind of filter is considerably narrower than those of the previously discussed designs. The disadvantage of this approach, however, is that numerous (unwanted) transmission spikes are present, the largest of which has the amplitude of about 75%.

It is obvious from these results that there are virtually an unlimited number of designs that can be tried, with corresponding tradeoffs in central passband width, stopband transmittance, presence of spikes in the stopband, and complexity of the birefringent filter. For all of these birefringent filter designs, tuning is continuous and easily implemented through rotation of the multiple crystals.

4.5 Laser Applications

When lasers were invented in 1960, they were called "a solution looking for a problem". Since then, they have become ubiquitous, finding utility in thousands of highly varied applications in every field of modern society, including consumer electronics, information technology, science, medicine, industry, law enforcement, entertainment, and military, etc.. In this section, we will briefly introduce laser applications in four fields: military, medical treatment, industry, and laser cooling.

4.5.1 Laser in Military

Military uses of lasers include applications such as lidar and rangefinder, target designation and ranging, defensive countermeasures, communications and directed energy weapons. Directed energy weapons are also in use, such as Boeing's Airborne Laser which is constructed inside a Boeing 747. It disrupts the trajectory of shoulder-fired missiles.

● **Lidar**

Light detection and ranging (LIDAR, also LADAR) is an optical remote sensing technology that can measure the distance to, or other properties of a target by illuminating the target with light, often using pulses from a laser. LIDAR technology has application in archaeology, geography, geology, geomorphology, seismology, forestry, remote sensing and atmospheric physics, as well as in airborne laser swath mapping (ALSM), laser altimetry and LIDAR contour mapping.

LIDAR has been used extensively for atmospheric research and meteorology. Downward-looking LIDAR instruments fitted to aircraft and satellites are used for surveying and mapping——a recent example being the National Aeronautics and Space Administration (NASA) experimental advanced research lidar. In addition, LIDAR has been identified by NASA as a key technology for enabling autonomous precision safe landing of future robotic and crewed lunar landing vehicles.

LIDAR uses ultraviolet, visible, or near infrared light to image objects and can detect a wide range of targets, including non-metallic objects, rocks, rain, chemical compounds, aerosols, clouds and even single molecules. A narrow laser beam can be used to map physical features with very high resolution.

In general, there are two kinds of lidar detection schema: incoherent or direct energy detection (which is principally an amplitude measurement) and coherent detection (which is best for doppler, or phase sensitive measurements). Coherent systems generally use optical heterodyne detection which is more sensitive than direct detection, that allows them to operate a much lower power but at the expense of more complex transceiver requirements.

In both coherent and incoherent LIDAR, there are two types of pulse models: micro pulse lidar systems and high energy systems. Micro pulse systems have developed as a result of the ever increasing amount of computer power available combined with advances in laser technology. They use considerably less energy in the laser, typically on the order of one micro joule, and are often "eye-safe", meaning they can be used without safety precautions. High-power systems are common in atmospheric research, where they are widely used for measuring many atmospheric parameters: the height, layering and densities of clouds, cloud particle properties (extinction coefficient, backscatter coefficient, depolarization), temperature, pressure, wind, humidity, trace gas concentration (ozone, methane, nitrous oxide, etc.).

There are several major components to a LIDAR system:

—Laser

600～1000nm lasers are most common for non-scientific applications. They are inexpensive, but since they can be focused and easily absorbed by the eye, the maximum power is limited by the need to make them eye-safe. Eye-safety is often a requirement for most applications.

—Scanner and optics

How fast images can be developed is also affected by the speed at which they are scanned. There are several options to scan the azimuth and elevation, including dual oscillating plane mirrors, a combination with a polygon mirror, a dual axis scanner. Optic choices affect the angular resolution and range that can be detected. A hole mirror or a beam splitter are options to collect a return signal.

—Photodetector and receiver electronics

Two main photodetector technologies are used in lidar: solid state photodetectors, such as silicon avalanche photodiodes, or photomultipliers. The sensitivity of the receiver is another parameter that has to be balanced in a LIDAR design.

—Position and navigation systems

LIDAR sensors that are mounted on mobile platforms such as airplanes or satellites require instrumentation to determine the absolute position and orientation of the sensor. Such devices generally include a Global Positioning System (GPS) receiver and an Inertial Measurement Unit (IMU).

● **Laser rangefinder**

A laser rangefinder is a device which uses a laser beam to determine the distance to an object. The most common form of laser rangefinder operates on the time of flight principle by sending a laser pulse in a narrow beam towards the object and measuring the time taken by the pulse to be reflected off the target and returned to the sender. Due to the high speed of light, this technique is not appropriate for high precision sub-millimeter measurements, where triangulation and other techniques are often used.

Rangefinders provide an exact distance to targets located beyond the distance of point-blank shooting to snipers and artillery. They also can be used for military reconciliation and engineering.

Handheld military rangefinders operate at ranges of 2km up to 25km and are combined with binoculars or monoculars. When the rangefinder is equipped with a digital magnetic compass (DMC) and inclinometer, it is capable of providing magnetic azimuth, inclination, and height (length) of targets. Some rangefinders can also measure the speed of the target in relation to the observer. Some rangefinders have cable or wireless interfaces to enable them to transfer their measurement data to other equipment like fire control computers. Some models also offer the possibility to use add-on night vision modules. Most handheld rangefinders use standard or rechargeable batteries.

The more powerful models of rangefinders measure distance up to 25km and are normally installed either on a tripod or directly on a vehicle or gun platform. In the latter case the rangefinder module is integrated with on-board thermal, night vision and daytime observation equipment. The most advanced military rangefinders can be integrated with computers.

To make laser rangefinders and laser-guided weapons less useful against military targets, various military arms may have developed laser-absorbing paint for their vehicles. Regardless, some objects don't reflect laser light very well and using a laser rangefinder on them is difficult.

● **Defensive countermeasures**

Defensive countermeasure applications can range from compact, low power infrared countermeasures to high power, airborne laser systems. IR countermeasure systems use lasers to confuse the seeker heads on heat-seeking anti-aircraft missiles. High power boost-phase intercept laser systems use a complex system of lasers to find, track and destroy intercontinental ballistic missiles (ICBM). In this type of system a chemical laser, in which the laser operation is powered by an energetic chemical reaction, is used as the main weapon beam. The mobile tactical high-energy laser (MTHEL) is another defensive laser system under development; this is envisioned as a field-deployable weapon system able to track incoming artillery projectiles and cruise missiles by radar and destroy them with a powerful deuterium fluoride laser.

Another example of direct use of a laser as a defensive weapon was researched for the Strategic Defense Initiative (SDI, nicknamed "Star Wars"), and its successor programs. This project would use ground-based or space-based laser systems to destroy incoming ICBMs. The practical problems of using and aiming these systems were many; particularly the problem of destroying ICBMs at the most opportune moment, the boost phase just after launch. This would involve directing a laser through a large distance in the atmosphere, which, due to optical scattering and refraction, would bend and distort the laser beam, complicating the aiming of the laser and reducing its efficiency.

Another idea from the SDI project was the nuclear-pumped X-ray laser. This was essentially an orbiting atomic bomb, surrounded by laser media in the form of glass rods; when the bomb exploded, the rods would be bombarded with highly-energetic gamma-ray photons, causing spontaneous and stimulated emission of X-ray photons in the atoms making up the rods. This would lead to optical amplification of the X-ray photons, producing an X-ray laser beam that would be minimally affected by atmospheric distortion and capable of destroying ICBMs in flight. The X-ray laser would be a strictly one-shot device, destroying itself on activation. Some initial tests of this concept were performed with underground nuclear testing; however, the results were not encouraging. Research into this approach to missile defense was discontinued after the SDI program was cancelled.

- **Target designator**

Another military use of lasers is as a laser target designator. This is a low-power laser pointer used to indicate a target for a precision-guided munition, typically launched from an aircraft. The guided munition adjusts its flight-path to home in to the laser light reflected by the target, enabling a great precision in aiming. The beam of the laser target designator is set to a pulse rate that matches that set on the guided munition to ensure munitions strike their designated targets and do not follow other laser beams which may be in use in the area. The laser designator can be shined onto the target by an aircraft or nearby infantry. Lasers used for this purpose are usually infrared lasers, so the enemy cannot easily detect the guiding laser light.

- **Laser sight**

Laser has been used in most firearms applications as a tool to enhance the targeting of other weapon systems. For example, a laser sight is a small, usually visible-light laser placed on a handgun or a rifle and aligned to emit a beam parallel to the barrel. Since a laser beam has low divergence, the laser light appears as a small spot even at long distances; the user places the spot on the desired target and the barrel of the gun is aligned (but not necessarily allowing for bullet drop, windage and the target moving while the bullet travels).

Most laser sights use a red laser diode. Others use an infrared diode to produce a dot invisible to the naked human eye but detectable with night vision devices. In the late 1990s, green diode pumped solid state laser (DPSS) laser sights (532 nm) became available. Modern laser sights are small and light enough for attachment to the firearms.

In 2007, Laser Max, a company specializing in manufacturing lasers for military and police firearms, introduced the first mass-production green laser available for small arms. This laser mounts to the underside of a handgun or long arm on the accessory rail. The green laser is supposed to be more visible than the red laser in bright lighting conditions because, for the same wattage, green light appears brighter than red light.

- **Eye-targeted lasers**

A non-lethal laser weapon was developed by the U.S. Air Force to temporarily impair an adversary's ability to fire a weapon or to otherwise threaten enemy forces. This unit illuminates an opponent with harmless low-power laser light and can have the effect of dazzling or disorienting the subject or causing him to flee. Several types of dazzlers are now available, and some have been used in combat.

There remains the possibility of using lasers to blind, since this requires much lower power levels, and is easily achievable in a man-portable unit. However, most nations regard the deliberate permanent blinding of the enemy as forbidden by the rules of war. Although several nations have developed blinding laser weapons, such as China's ZM-87, none of these are believed to have made it past the prototype stage.

In addition to the applications that crossover with military applications, a widely known law enforcement use of lasers is for lidar to measure the speed of vehicles.

4.5.2 Laser in Medicine

Medicine and the biosciences are important fields for implementing lasers. Even before early lasers had left research laboratories for industrial applications, medicine was considered an important "consumer" of laser technologies, first as manufacturing tools for medical instruments, then as working instruments themselves. Practically all types of lasers have found their specific niches in important branches of medicine: research, monitoring, imaging, probing, therapy, surgery, and others. Referring to more specific applications, lasers are used literally everywhere. In biomedical investigation: fluorescent spectroscopy, microscopy, and flow cytometry. In surgery: "bloodless" operations in cardiology, on abdominal and thoracic organs, and skull and brain microsurgery. In cosmetic and aesthetic medicine: smoothing wrinkles, resurfacing the skin, and bleaching tattoos. In therapy: the treatment of cancer, spider veins, and vascular dysfunction. In diagnostics: endoscopic investigations and optical coherence tomography (OCT). This list can be extended further by going deeper into subclassifications and interdisciplinary topics.

Depending on the particular requirements, numerous types of lasers can provide different wavelengths, energy levels, and operation modes. As a short overview, Table 4.1 provides some examples of the most typical and commercially available systems, as well as their particular uses.

Table 4.1 Main Laser Types and Fields of Applications in Medicine

Laser type and operation mode	Wavelength (μm)	Application
Carbon dioxide(CW, pulsed)	10.6	Surgery: general and eye; dental therapy
Argon(CW)	0.488, 0.514	Sealing blood vessels in retina, eye microsurgery, plastic surgery, photodynamic therapy
Nd:YAG(CW, Q-switched)	1.06	General surgery, dentistry: therapy and surgery
Nd:YAG(Q-switched)	0.532 (double frequency)	Surgery, ophthalmology, dermatology, cosmetic, photodynamic therapy
Ruby(Q-switched)	0.694	Plastic surgery, dermatology, photodynamic therapy
Er:YAG(Q-switched)	2.94	Skin resurfacing (superficial ablation), dental therapy and surgery
Ho:YAG(CW, pulsed)	2.12	Ablation, incision, tissue hemostatic vaporization, cancer tumor treatment
Diode lasers(CW, pulsed)	0.63, 0.82, 0.83, 0.98,1.45	Photodynamic therapy, endovenous treatment, aesthetic medicine, vascular lesions
Alexandrite lasers (CW, Q-switched)	0.755	Pigmented lesions, tattoo bleaching, vascular lesions, skin treatment, hair removal
Dye lasers (CW, Q-switched, tunable)	0.570～0.650	Treatment of malignant tissues, photodynamic therapy, cosmetics, vascular lesions, hair removal

Specially, fiber lasers should be considered successors of trends in medical applications rather than "pioneers" discovering untouched fields. However, due to their inherent flexibility of physical principles and design, as well as outstanding performance, fiber lasers have enormous potential to bring new opportunities to medicine.

Before specifying particular applications of fiber lasers in medicine, it is instructive to briefly depict some of the main fields in which lasers are commonly used for health care, monitoring, and research regardless of which particular type of laser is considered. Such a classification is commonly arranged according to how organic tissues react to laser radiation.

Using the terminology adopted among medical professionals, the reaction of organic tissues to laser radiation is typically described as an optical or thermal response, as shown in Fig.4.14.

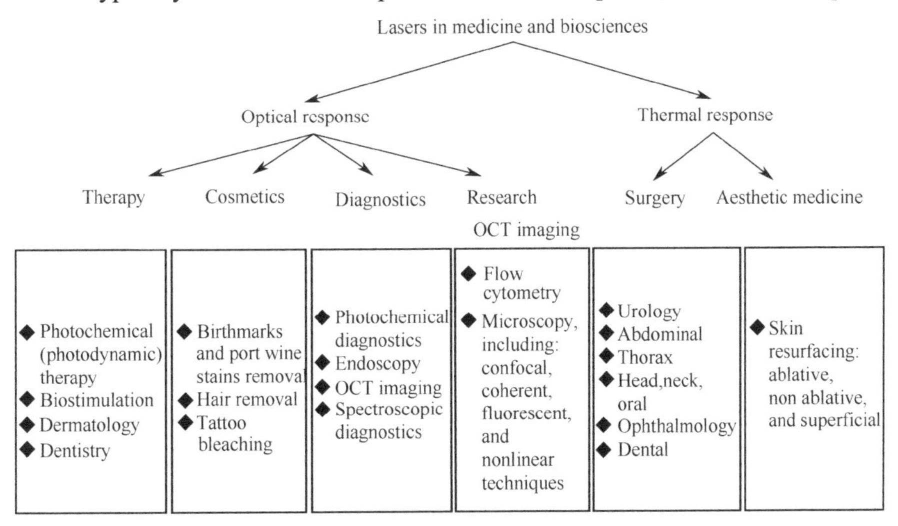

Fig.4.14 General applications of laser in medicine and life sciences

In the optical response, the light energy absorbed does not damage or destroy the tissues. Most effects are achieved either (1) by selective resonance absorption of specific laser wavelengths by fluorophores or photosensitizers with sequentially photoinduced changes of the tissues, or (2) by exposure with short light pulses of high peak intensity leading to material ablation. Thermal response is normally produced by CW or long-pulse laser radiation, when larger power delivered to organs is converted into heat and destroys the surrounding tissue. How tissues react to laser radiation in particular depends on the chosen wavelength, mode of operation, pulse duration, and energy, as well as the laser spot size.

4.5.3 Laser in Industry

Industry is another important field for implementing laser. Here we just provide some aspects of its applications.

- **Laser cutting**

Laser cutting is a technology that uses a laser to cut materials, and is typically used for industrial manufacturing applications, but is also starting to be used by schools, small businesses and hobbyists. Laser cutting works by directing the output of a high-power laser, by computer, at the material to be cut. The material then either melts, burns, vaporizes away, or is blown away by a jet of gas, leaving an edge with a high-quality surface finish.

Its advantages include easier work holding and reduced contamination of workpiece. Precision may be better since the laser beam does not wear during the process. There is also a reduced chance of warping the material that is being cut, as laser systems have a small heat-affected zone.

- **Laser peening**

Laser peening, or laser shock peening (LSP), is a process of hardening or peening metal using a powerful laser. Laser peening can impart a layer of residual compressive stress on a surface that is four times deeper than attainable from conventional shot peening treatments. A coating, usually black tape or paint, is applied to absorb the energy. Short energy pulses are then focused to explode the ablative coating, producing a shock wave. The beam is then repositioned and the process is repeated, creating an array of slight indents of compression and depth with about 5%～7% cold work. A translucent layer, usually consisting of water, is required over the coating and acts as a tamp, directing the shock wave into the treated material. This computer-controlled process is then repeated, often as many as three times, until the desired compression level is reached, producing a compressive layer as deep as 1～2mm average.

Laser peening is often used to improve the fatigue resistance of highly stressed critical turbine engine components, and the laser (or component) is typically manipulated by an industrial robot.

- **Laser graving**

Laser graving, or laser marking, is the practice of using lasers to engrave or mark an object. The technique does not involve the use of inks, nor does it involve tool bits which contact the engraving surface and wear out. These properties distinguish laser engraving from alternative engraving or marking technologies where bit heads have to be replaced regularly or inks have to be used.

The impact of laser engraving has been more pronounced for specially-designed "laserable" materials. These include laser-sensitive polymers and novel metal alloys.

The term laser marking is also used as a generic term covering a broad spectrum of surfacing techniques including printing, hot-branding and laser bonding. The machines for laser engraving and laser marking are the same, so that the two terms are usually exchangeable.

- **Laser bonding**

Laser bonding is a marking technique that uses lasers and other forms of radiant energy to bond an additive marking substance to a wide range of substrates.

Invented in 1997, this patent protected technology delivers permanent marks on metals, glass and ceramic parts for a diverse range of industrial and manual applications, ranging from aerospace to the awards & engraving industries. It differs from the more widely known techniques of laser engraving and laser ablation in that it is an additive process, adding material to the substrate to form the marking instead of removing it as in those techniques.

4.5.4 Laser Cooling

A technique that has recent success is laser cooling. This involves atom trapping, a method where a number of atoms are confined in a specially shaped arrangement of electric and magnetic fields. Shining particular wavelengths of laser light at the ions or atoms slows them down, thus cooling them. As this process is continued, they all are slowed and have the same energy level, forming an unusual arrangement of matter known as a Bose-Einstein condensate.

Laser cooling refers to the number of techniques in which atomic and molecular samples are cooled through the interaction with one or more laser light fields. The first example of laser cooling, and also still the most common method of laser cooling (so much so that it is still often referred to as "laser cooling") is Doppler cooling.

- **Doppler cooling**

Doppler cooling, which is usually accompanied by a magnetic trapping force to give a magneto-optical trap, is by far the most common method of laser cooling. It is used to cool low density gasses down to the Doppler cooling limit, which for Rubidium 85 is around 150 microkelvin. As Doppler cooling requires a very particular energy level structure, known as a closed optical loop, the method is limited to a small handful of elements.

In Doppler cooling, the frequency of light is tuned slightly below an electronic transition in the atom. Because the light is detuned to the "red" (i.e. at lower frequency) of the transition, the atoms will absorb more photons if they move towards the light source, due to the Doppler effect. Thus if one applies light from two opposite directions, the atoms will always scatter more photons from the laser beam pointing opposite to their direction of motion. In each scattering event the atom loses a momentum equal to the momentum of the photon. If the atom, which is now in the excited state, emits a photon spontaneously, it will be kicked by the same amount of momentum but in a random direction. The result of the absorption and emission process is to reduce the speed of the atom, provided its initial speed is larger than the recoil velocity from scattering a single photon. If the absorption and emission are repeated many times, the mean velocity, and therefore the kinetic energy of the atom will be reduced. Since the temperature of an ensemble of atoms is a measure of the random internal kinetic energy, this is equivalent to cooling the atoms.

- **Other methods of laser cooling**

Several somewhat similar processes are also referred to as laser cooling, in which photons are used to pump heat away from a material and thus cool it. The phenomenon has been demonstrated via anti-Stokes fluorescence, and both electroluminescent upconversion and photoluminescent upconversion have been studied as means to achieve the same effects. In many of these, the coherence of the laser light is not essential to the process, but lasers are typically used to achieve a high irradiance.

Laser cooling is primarily used for experiments in Quantum Physics to achieve temperatures of near absolute zero (−273.15℃, −459.67℉). This is done to observe the unique quantum effects that can only occur at this heat level. Generally, laser cooling has been only used on the atomic level to cool down elements. This may soon change, as a new breakthrough in the technology has successfully cooled a macro-scale object to near absolute zero.

4.5.5 Laser in Daily Life

Laser has sucuessfully underlined and comprised our daily life as an inevitably developlement of laser techniques.

- **Laser marking**

Using a laser to mark or code information on a product—laser marking is one of the most common applications of lasers. Laser marking often takes the form of an alphanumeric code imprinted the label or on the surface of the product to describe date of manufacture, best-before, serial number or part number, but the mark can also be a machine-readable bar code or 2D symbol (ID matrix). This type of marking is often specified as laser coding. Since the equipment used for surface processing of a macroscopic nature (e.g. scribing, melting, smoothing or roughening, annealing, etc) is basically the same equipment sold laser marking, the term laser marking has taken a more general meaning than simple coding. Hence, as well as coding, laser marking also includes functional marking (such as scribing gradation lines on a syringe) decorative marking (such as engraving a logo or graphic image on an integrated circuit).

- **Laser coding**

Laser coding is one of the final processes in the assembly of a product, taking place during the final filling cycle at a brewery, for example, or on a finished integrated-circuit chip before it is boxed for shipment. Compared to other on-line marking techniques such as Inkjet, hot stamping or mechanical scribing, laser coding offers many advantages: indelibility, reliability, no consumables and high speed. Laser coding is usually the best marking solution with one proviso: not all materials mark well with every laser.

- **Optical data storage**

Optical data storage is found in popular consumer products. Compact discs (CDs), digital versatile discs (DVDs) and minidiscs (MDs), are all forms of optical data storage. More advanced forms of optical data storage include high-speed devices and library products. All optical data storage devices use optical principles to achieve high data density, rugged packaging, reliable information retrieval and cost-effective production. In general, optical data storage relates to placing information on a surface so that, when a light beam scans the surface, the reflected light can be used to recover the information. There are many forms of optical storage media and many types of optical systems used to scan data. This chapter discusses the basic principles of optical data storage, types of commercial optical media available in 2002, several performance parameters and some interesting prospects for future systems.

- **Laser printing**

A laser printer uses a laser to form an electrostatic latent image on a photoconductor and then uses the electrophotographic process to visualize this image. The imaging process of a laser printer is basically the same as that used in conventional analogue photocopiers. Laser printers and digital photocopiers generally employ a reversal (negative-positive) developing process whereby areas of the photoconductor corresponding to black regions of the image are discharged by exposing them to light, and these discharged areas are then developed to make them visible.

Words and Expressions

abdominal	腹的
ablation	消融
accessory	附属的，附加的，辅助的
aerospace	航空宇宙
aesthetic	美学的
alexandrite	翠绿石
align	排列，匹配
alloy	合金
alphanumeric	字母数字的
amplifying rod	放大棒
anneal	退火，磨炼
angular divergence	发散角
arise from	由……引起，由……产生
artillery	炮兵，炮队
atomic bomb	原子弹
attainable	可达到的
avalanche	雪崩
azimuth	方位，方位角
barrel	枪管
beam waist	束腰
best-before	有效期
binocular	双目镜，双筒镜

birefringent	双折射
bleach	漂白
bonding	焊接
brewery	啤酒厂
but for	如果没有……
cardiology	心脏病学
cavity	腔
ceramic	陶器的，陶瓷的
coefficient	系数
combat	战斗
confocal	共焦
contamination	污染
cosmetic	美容的
consumer	用户
countermeasure	干扰，对抗
criterion	判据，标准
cruise missile	巡航导弹
curvature	曲率
cylindrical	圆柱形的
cytometry	血细胞计数
dazzle	使目眩，使眼花
depict	描绘
deplete	枯竭，消耗
deploy	展开，部署，使用
depopulation	粒子数减少
dermatology	皮肤病学
designator	指示器
diminish	减小
discriminate	识别，区分
disrupt	毁坏，干扰，中断
dual axis scanner	双轴扫描仪
dye	染料
electroluminescent	场致发光的
electrophotographic	电子照相的
electrostatic	静电的
elevation	高度，海拔
elongation	拉长，延长
endoscopic	内窥镜的
enforcement	强制执行
engrave	雕刻

ensemble	全体，总效果
envision	预见，展望
extensively	广大的
far-field divergence	远场发散角
fatigue	疲劳
flee	逃脱，逃避
flow cytometry	流式细胞术
fluorescence	荧光
forestry	林业学
fusion	聚变
gaussian profile	高斯包络，高斯分布
geology	地质学
geomorphology	地形学
gradation	渐变
graving	雕刻
hemispherical resonator	半球腔
heterodyne	外差法
heuristic	启发式的，探索的
humidity	湿度、湿气
hyperbola	双曲线
imprint	印记，特征
inclinometer	倾角仪，磁角仪
indelibility	不能消除
infantry	步兵，步兵团
inhibit	阻止，禁止
inkjet	喷墨, 喷墨打印机
in terms of	就……而言
intercontinental ballistic missiles	洲际弹道导弹
interferometer	干涉仪
irradiance	辐照度，发光
kinetic energy	动能
laser	激光
laser sight	激光瞄准
lethal	致命（性）的
lidar	激光雷达
longitudinal	纵向的
lunar landing	登月
machine-readable	机器可读的，可用计算机处理的
macroscopic	宏观的
micro joule	微焦耳

minidisc	小型磁盘
miniature	小型的，微小的
monitor	监控
munition	弹药，军火
navigation	导航
niche	适当的位置
non-lethal	非致命的
ophthalmology	眼科学
organic	有机的
oscillation	振荡
out of phase	异相
parentheses	括号
peening	锤击，敲打
photocopier	复印机
photodetector	光电探测器
photodiode	二极管
photosensitizer	光敏剂
photomultiplier	光电倍增管
photon	光子
piezoelectric	压电的
polymer	聚合物
polynomial	多项式
portable	便携式的，手提式的
portion	部分
precautions	防范、预防措施
preferentially	优先地
probe	探测
proviso	附带条件
rangefinder	测距仪
recoil	反冲，返回
reconciliation	和解
reference	参考
resonance	共振，谐振
retain	保持
retardation	延迟，光程差
rifle	步枪，来复枪
round-trip	往返，双程
rubidium	铷
rugged	高低不平的，粗糙的
schema	计划

seismology	地震学
servo system	伺服系统
skull	颅骨
smoothing	平滑化
sniper	狙击手
spectral range	光谱响应范围
spider veins	蛛状曲张静脉
splitter	分束器
spontaneous emission	自发辐射
stringent	严格的，精确的
syringe	注射器
tactical	战术的，作战的
tamp	夯具
tattoo	纹身
terminate	终止
therapy	治疗，疗法
thermal expansion	热膨胀
thoracic	胸的
thus far	迄今，到此为止
tissue	组织
tradeoff	折衷（办法、方案），综合，权衡
trajectory	弹道，轨迹，轨道
translucent	半透明的
tripod	三脚架
tunable	可调的
turbine	涡轮（机）
ubiquitous	普遍存在的，无所不在的
ultraviolet	紫外线（UV）
unstable	非稳定的
utility	实用、效用
vascular dysfunction	血管功能障碍
vice visa	反之亦然
vignette	渐晕
warp	变形
wattage	瓦特

Grammar　专业英语翻译方法（四）：定语从句的译法

定语从句在主从复合句中起定语作用。引导定语从句的关联词有：

关系代词：who，whom，whose，which，that，as，but

关系副词：when，where，how，why

关系代词或关系副词在先行词和定语从句之间起纽带作用，同时又作为定语从句的一个句子成分。关系代词用作定语，关系副词用作状语。

—Nothing in the world is difficult for one who sets his mind to it.

世上无难事，只怕有心人。

—The quantity of heat which is supplied to the engine can be found from the weight of fuel oil which is consumed.

供给发电机的热量可根据消耗掉的燃油质量求得。

—The upper part of the cylinder liners where high combustion temperatures are met with is water cooled.

与燃烧高温相接触的气缸套上部是水冷却的。

—He told me how he had read about it in The People's Daily.

他告诉我，他是在《人民日报》上读到这件事的。

—There is another reason why A.C. is preferred to D.C. for long-distance transmission.

为什么长途输电时交流电优于直流电，还有另一个理由。

定语从句可分为限制性定语从句和非限制性定语从句。限制性定语从句是先行词不可缺少的定语，等于是给它下定义，成为非一般性的词意。翻译时一般把从句放在它所修饰词之前。非限制性定语从句是先行词的附加说明。翻译时一般可以顺译，把定语从句译成分句或句子的其他成分。

（1）对于限制性定语从句一般采用合译法，译成“……的”，但整个句子的语序一般要破译。

—The volt is the unit which is used for measuring potential.

伏特是用来测量电位的单位。

限制性定语从句也可以合译成其他句子成分。

—Good clocks have pendulums which are automatically compensated for temperature changes.

好钟的摆可以自动补偿温度变化。

（2）对于非限制性定语从句一般采用分译法，即将定语从句与主句分开翻译，根据定语从句的含义译成并列句或状语从句。当定语从句说明的不是某一个先行词而是整个主句时，也可译为并列分句，语序一般可顺译。

（a）当非限制性定语从句用以补充说明其前置词时，可译成并列分句而将前置词重复或以“它”和其他代词代替，或在其前加“这”或“该”，有时也可将关系代词 which 省略不译。

—Sliver is a good conductor, which allows electric current to flow easily.

银是良导体，它能让电流容易通过。

—An element is a substance, which cannot be broken down into smaller substance by ordinary chemical means.

元素是一种物质，这种物质不能用一般的化学方法再分裂成更简单的物质。

（b）当非限制性定语从句用以说明整个主句时，仍可译成并列句，其关系代词可译成指示代词。

—Friction wears away metal in the moving parts, which shortens their working life.

运动零件间的摩擦力使金属磨损，这就缩短了它们的使用寿命。

—Miniaturization means making things small in size, which is of great importance for the development of the electronics industry.

小型化意味着缩小东西的尺寸，这对电子工业的发展极为重要。

（c）有些定语从句跟主句在逻辑上有状语关系，实际上起主句的原因、结果、目的、条件、让步等状语的作用，可译成状语从句，但要添加适当的从属连词。

—Air will completely fill any container, in which it may be placed.

不管空气装在什么容器里，它都能把容器充满。

（d）把非限制性定语从句做其他性质的句子成分。

—Hydrogen, which is the lightest element, has only electron.

最轻的元素氢只有一个电子（同位语）。

（3）关联词前带介词的定语从句一般为限制性的（“of which”），也可以是非限制性的，翻译时需考虑关联词前面的介词的含义。

—Electronic computer is a subject about which many books have written.

电子计算机是许多书所围绕的一个主题。

—The neutrons and protons combine the nucleus of the atom, around which the electrons move.

中子和质子组成原子核，而电子则围绕原子核旋转。

Part 5 Optical Fiber Communication

In an optical communication system, information is transmitted by light propagation inside an optical fiber in the form of a coded sequence of optical pulses. The signal weakens during propagation because of fiber loss. After some distance it becomes necessary to regenerate the signal through the use of a repeater (essentially a detector-amplifier-transmitter combination). System cost considerations require that the repeater spacing L, be as large as possible. Another consideration is related to the transmission capacity as determined by the bit rate, B (number of bits transmitted per second). The objective of many lightwave transmission systems is to maximize the bit rate-distance product, BL. The bit rate is limited inherently by chromatic dispersion in the fiber, which is responsible for the broadening of optical pulses during their propagation inside a single-mode fiber. The choice of the operating wavelength is therefore related to the loss and dispersion characteristics of the fiber.

5.1 Development of Optical Communication System

The first generation of lightwave transmission systems utilized GaAs lasers operating at a wavelength of about 0.85μm. Relatively high values of the loss and dispersion coefficients restricted the repeater spacing to ~10km and the bit rate to ~100Mb/s.

The second generation of systems made use of the wavelength region around 1.3μm, where fiber dispersion is negligible. The use of InGaAs lasers coupled with the relatively low fiber loss allowed a repeater spacing of about 20km. However, the bit rate had to be bellow ~100Mb/s because of modal dispersion in multimode fibers. This problem was overcome with the use of single-mode fibers; the absence of chromatic dispersion near 1.3μm then allowed much higher bit rates (up to 2Gb/s). However, the repeater spacing (~50km) was limited by the fiber loss at this wavelength.

Minimum fiber loss occurs around 1.55μm. The third generation of optical communication systems is therefore based on 1.55μm InGaAs lasers. At this wavelength the repeater spacing can easily exceed 100km for moderate bit rates. At high bit rates (B >1Gb/s) the repeater spacing is limited not by the fiber loss but rather by the extent of fiber dispersion. Two distinct routes are being followed to overcome this problem. In one approach the zero-dispersion wavelength, which is about 1.3μm for conventional silica fibers, is shifted towards the desirable 1.55μm region by modifying the fiber characteristics. In the other approach the effect of fiber dispersion is minimized by reducing the spectral width of the 1.55μm InGaAs laser source.

The fourth-generation of optical communication systems makes use of homodyne or heterodyne detection techniques. Because of the phase-sensitive nature of such systems, they are referred to as coherent communication system. For the same reason, they require tunable semiconductor lasers with a narrow linewidth (usually <10MHz or 10^{-4}nm). Special multisection lasers have been developed to meet the demands of fourth-generation lightwave systems.

The fifth generation of lightwave systems employs erbium-doped fiber amplifiers for amplifying the transmitted optical signal periodically to compensate for fiber loss. The transmission distance can exceed a few thousand kilometers for such systems as long as fiber dispersion does not limit it. A novel approach makes use of optical solitons that use fiber nonlinearity for compensating fiber dispersion. Such solitons can travel along the fiber for thousands of kilometers without experiencing temporal broadening as long as fiber loss is compensated through periodic amplification. The development of such fifth-generation lightwave systems requires new kinds of semiconductor lasers. The use of optical solitons requires semiconductor lasers capable of producing ultrashort optical pulses (pulse width < 50ps) at high repetition rates. Advances in the semiconductor-laser technology play an important role in realizing such high-performance lightwave systems.

5.2 Optical Fiber Characteristics

In its simplest form, an optical fiber consists of a central core surrounded by a cladding layer whose refractive index is slightly lower than that of the core. Such fibers are generally referred to as step-index fibers to distinguish them from graded-index fibers in which the refractive index of the core decreases gradually from center to the core boundary. Fig.5.1 shows schematically the cross section and the refractive-index profile of a step-index fiber. Two parameters which characterize the fiber are the relative core-cladding index difference Δ defined by

$$\Delta = \frac{n_1 - n_2}{n_1} \tag{5.1}$$

and the normalized frequency V defined by

$$V = k_0 a (n_1^2 - n_2^2)^{1/2} \tag{5.2}$$

where $k_0 = 2\pi/\lambda$, a is the core radius, and λ is the wavelength of light.

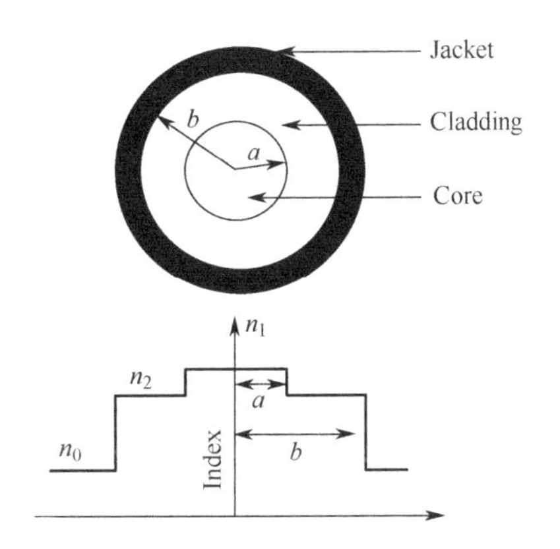

Fig.5.1 Schematic illustration of the cross section and the refraction-index profile of a step-index fiber

The parameter V determines the number of mode supported by the fiber. A step-index fiber supports a single mode if $V < 2.405$. The fibers designed to satisfy this condition are called single-mode fibers. The main difference between the single-mode and multimode fibers is the core size. The core radius $a = 25\sim30\mu m$ for typical multimode fibers. However, single-mode fibers with a typical value of $\Delta \approx 3\times10^{-3}$ require a to be in the range 2~4μm. The numerical value of the outer radius b is less critical as long as it is large enough to confine the fiber modes entirely. Typically b = 50~60μm for both single-mode and multimode fibers.

5.2.1 Optical Losses

An important fiber parameter is a measure of power loss during transmission of optical signals inside the fiber. If P_0 is the power launched at the input of a fiber of length L, the transmitted power P_T is given by

$$P_T = P_0 \exp(-\alpha L) \tag{5.3}$$

where α is the attenuation constant, commonly referred to as the fiber loss. It is customary to express the fiber loss in units of dB/km by using the relation

$$\alpha_{\text{dB}} = -\frac{10}{L}\log\left(\frac{P_T}{P_0}\right) = 4.343\alpha \tag{5.4}$$

where Eq.5.4 was used to relate α_{dB} and α.

The fiber loss depends on the wavelength of light. Fig.5.2 shows the loss spectrum of a silica fiber. This fiber exhibits a minimum loss of about 0.2dB/km near 1.55μm. The loss is considerably higher at shorter wavelengths, reaching a level of 1~10dB/km in the visible region. Note, however, that even a 10dB/km loss corresponds to an attenuation constant of only $\alpha \approx 2\times10^{-5}\text{cm}^{-1}$.

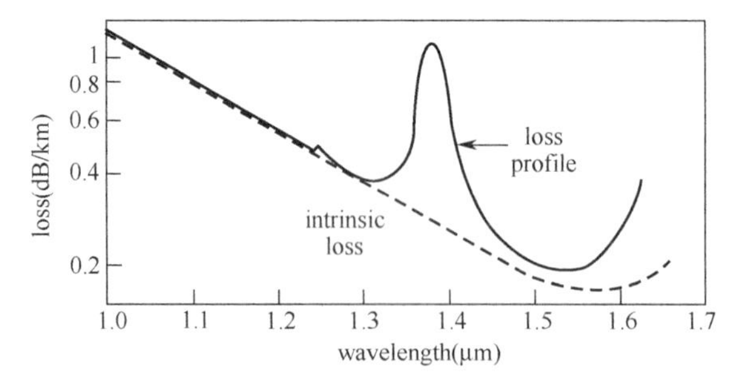

Fig.5.2 Measured loss spectrum of a single-mode silica fiber

Several factors contribute to the loss spectrum, material absorption and Rayleigh scattering contributing dominantly. Other factors that may contribute to the fiber loss are bending losses and boundary losses (due to scattering at the core-cladding boundary). The total loss of a fiber link in the optical communication systems also includes the splice losses that occurs when two fiber pieces are joined together. Advances in fiber technology have reduced splice losses to a level~0.01dB/km.

5.2.2 Chromatic Dispersion

When an electromagnetic wave interacts with bound electrons of a dielectric, the medium response in general depends on the optical frequency ω. This property, referred to as chromatic dispersion, manifests through the frequency dependence of the refractive index $n(\omega)$. On a fundamental level, the origin of chromatic dispersion is related to the characteristic resonance frequencies at which the medium absorbs the electromagnetic radiation through oscillations of bound electrons. Far from the medium resonances, the refractive index is well approximated to the Sellmeier equation

$$n^2(\omega) = 1 + \sum_{j=1}^{m} \frac{B_j \omega_j^2}{\omega_j^2 - \omega^2} \tag{5.5}$$

where ω_j is the resonance frequency and B_j is the strength of jth resonance. The sum in Eq.(5.5) extends over all material resonances that contribute to the frequency range of interest. In the case of optical fibers, the parameters B_j and ω_j are obtained experimentally by fitting the measured dispersion curve to Eq.(5.5) with $m = 3$ and depend on the core constituent.

Fiber dispersion plays a critical role in propagation of short optical pulses since different spectral components associated with the pulse travel at different speeds given by $c/n(\omega)$. Even when nonlinear effects are not important, dispersion-induced pulse broadening can be detrimental for optical communication systems.

Mathematically, the effects of fiber dispersion are accounted for by expanding the mode-propagation constant β in a Taylor series about the frequency ω_0 at which the pulse spectrum is centered:

$$\beta(\omega) = n(\omega)\frac{\omega}{c} = \beta_0 + \beta_1(\omega - \omega_0) + \frac{1}{2}\beta_2(\omega - \omega_0)^2 + \cdots \tag{5.6}$$

where

$$\beta_m = \left.\frac{\mathrm{d}^m \beta}{\mathrm{d}\omega^m}\right|_{\omega=\omega_0} \qquad m = 0, 1, 2, \cdots \tag{5.7}$$

The parameters β_1 and β_2 are related to the refractive index n and its derivatives through the relations

$$\beta_1 = \frac{1}{v_g} = \frac{n_g}{c} = \frac{1}{c}\left(n + \omega\frac{\mathrm{d}n}{\mathrm{d}\omega}\right) \tag{5.8}$$

$$\beta_2 = \frac{1}{c}\left(2\frac{\mathrm{d}n}{\mathrm{d}\omega} + \omega\frac{\mathrm{d}^2 n}{\mathrm{d}\omega^2}\right) \tag{5.9}$$

where n_g is the group index and v_g is the group velocity. Physically speaking, the envelope of an optical pulse moves at the group velocity while the parameter β_2 represents dispersion of the group velocity and is responsible for pulse broadening. This phenomenon is known as the group-velocity dispersion (GVD), and β_2 is the GVD parameter. The coefficient β_3 appearing in that term is called the third-order dispersion (TOD) parameter. Such higher-order dispersive effects can distort ultrashort optical pulses both in the linear and nonlinear regimes.

Figures 5.3 and 5.4 show how n, n_g, and β_2 vary with wavelength λ in fused silica. The most notable feature is that β_2 vanishes at a wavelength of about 1.3 μm and becomes negative for longer wavelengths. This wavelength is referred to as the zero-dispersion wavelength and is denoted as λ_D.

It is possible to design dispersion-flattened optical fibers having low dispersion over a relatively large wavelength range 1.3~1.6μm. This can be achieved by using multiple cladding layers.

Nonlinear effects in optical fibers can manifest qualitatively different behaviors depending on the sign of the GVD parameter. For wavelengths such that $\lambda < \lambda_D$, the fiber is said to exhibit normal dispersion as $\beta_2 > 0$. In the normal-dispersion regime, high-frequency (blue-shifted) components of an optical pulse travel slower than low-frequency (red-shifted) components of the same pulse. By contrast, the opposite occurs in the anomalous dispersion regime in which $\beta_2<0$. As seen in Fig.5.4, silica fibers exhibit anomalous dispersion when the light wavelength exceeds the zero-dispersion wavelength ($\lambda > \lambda_D$). The anomalous-dispersion regime is of considerable interest for the study of nonlinear effects because it is in this regime that optical fibers support solitons through a balance between the dispersive and nonlinear effects.

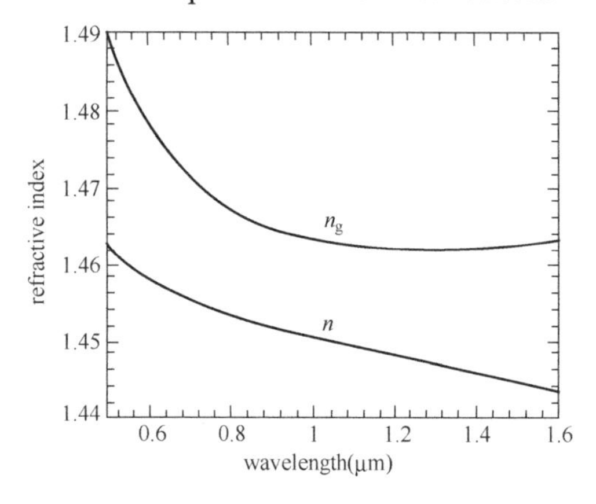

Fig.5.3 Variation of refractive index n and group index n_g with wavelength for fused silica

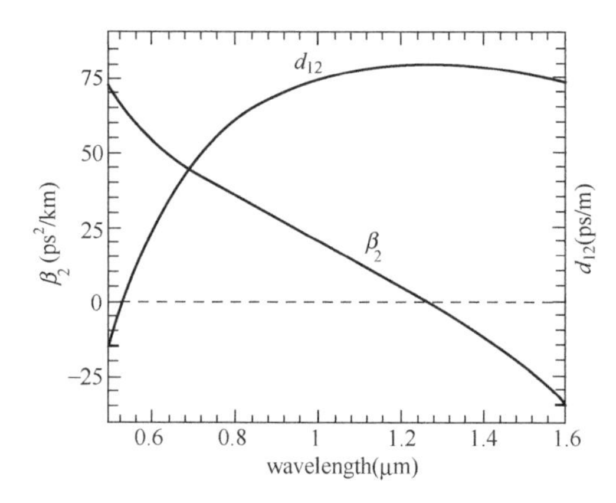

Fig.5.4 Variation of β_2 with wavelength for fused silica

An important feature of chromatic dispersion is that pulses at different wavelengths propagate at different speeds inside the fiber because of the group-velocity mismatch. This feature leads to a walk-off effect that plays an important role in the description of the nonlinear phenomena involving two or more overlapping optical pulses. More specifically, the nonlinear interaction between two optical pulses ceases to occur when the faster moving pulse has completely walked through the slower moving pulse. The group-velocity mismatch plays an important role in the case of the nonlinear effects involving cross-phase modulation.

5.2.3 Fiber Nonlinearities

The response of any dielectric to light becomes nonlinear for intense electromagnetic fields, and optical fibers are no exception. On a fundamental level, the origin of nonlinear response is related to a harmonic motion of bound electrons under the influence of an applied field. As a result, the total polarization $\boldsymbol{P}$ induced by electric dipoles is not linear in the electric field $\boldsymbol{E}$, but satisfies the more general relation

$$\boldsymbol{P} = \varepsilon_0[\chi^{(1)}\boldsymbol{E} + \chi^{(2)}\boldsymbol{E}^2 + \chi^{(3)}\boldsymbol{E}^3 + \cdots + \chi^{(n)}\boldsymbol{E}^n] \tag{5.10}$$

where ε_0 is the vacuum permittivity and $\chi^{(j)}$ $(j = 1,2,\cdots)$ is jth order susceptibility. In general, $\chi^{(j)}$ is a tensor of rank j+1. The linear susceptibility $\chi^{(1)}$ represents the dominant contribution to $\boldsymbol{P}$. Its effects are included through the refractive index n and the attenuation coefficient α. The second-order susceptibility $\chi^{(2)}$ is responsible for such nonlinear effects as second-harmonic generation and sum-frequency generation. However, it is nonzero only for media that lack of an inversion symmetry at the molecular level. As SiO_2 is a symmetric molecule, $\chi^{(2)}$ vanishes for silica glasses. As a result, optical fibers do not normally exhibit second-order nonlinear effects. Nonetheless, defects or color centers inside the fiber core can contribute to second-harmonic generation under certain conditions.

The lowest-order nonlinear effects in optical fibers originate from the third-order susceptibility $\chi^{(3)}$, which is responsible for phenomena such as third-harmonic generation, four-wave mixing, and nonlinear refraction. However, unless special efforts are made to achieve phase matching, the nonlinear processes which involve the generation of new frequencies are not efficient in optical fibers. Most of the nonlinear effects in optical fibers therefore originate from nonlinear refraction, a phenomenon that refers to the intensity dependence of the refractive index resulting from the contribution of $\chi^{(3)}$. The intensity dependence of the refractive index leads to a large number of interesting nonlinear effects; the two most widely studied are self-phase modulation (SPM) and cross-phase modulation (XPM). SPM refers to the self-induced phase shift experienced by an optical field during its propagation in optical fibers. SPM is responsible for spectral broadening of ultrashort pluses and the existence of optical solitons in the anomalous-dispersion regime of fiber. XPM refers to the nonlinear phase shift of an optical field induced by a copropagating field at a different wavelength. XPM is responsible for asymmetric spectral broadening of copropagating optical pulses.

The nonlinear effects governed by the third-order susceptibility $\chi^{(3)}$ are elastic in the sense that no energy is exchanged between the electromagnetic field and the dielectric medium. A second class of nonlinear effects results from stimulated inelastic scattering in which the optical field transfers part of its energy to the nonlinear medium. Two important nonlinear effects in optical fibers fall in this category; both of them are related to vibrational excitation modes of silica. These phenomena, known as stimulated Raman scattering (SRS) and stimulated Brillouin scattering (SBS), are among the first nonlinear effects studied in optical fibers. The main difference between the two is that optical phonons participate in SRS while acoustic phonons participate in SBS.

In a simple quantum-mechanical picture applicable to both SRS and SBS, a photon of the incident field (called the pump) is annihilated to create a photon at a lower frequency (belonging to the Stokes wave) and a phonon with the right energy and momentum to conserve the energy and the momentum. Of course, a higher-energy photon at the so-called anti-Stokes frequency can also be created if a phonon of right energy and momentum is available. Even though SRS and SBS are very similar in their origin, different dispersion relations for acoustic and optical phonons lead to some basic differences between the two. A fundamental difference is that SBS in optical fibers occurs only in the backward direction whereas SRS can occur in both directions.

5.3 Propagation of Optical Beam in Fiber

For an understanding of the nonlinear phenomena in optical fibers, it is necessary to consider the theory of electromagnetic wave propagation in dispersive nonlinear media.

Maxwell's equations can be used to obtain the wave equation that describes light propagation in optical fibers, i.e.,

$$\nabla^2 \widetilde{\boldsymbol{E}}(\boldsymbol{r},\omega)+\varepsilon(\omega)k_0^2\widetilde{\boldsymbol{E}}(\boldsymbol{r},\omega)=0 \tag{5.11}$$

where $k_0 = \omega / c$, $\widetilde{\boldsymbol{E}}(\boldsymbol{r},\omega)$ is the Fourier transform of $\boldsymbol{E}(\boldsymbol{r}, t)$ defined as

$$\widetilde{\boldsymbol{E}}(\boldsymbol{r},\omega) = \int_{-\infty}^{+\infty} \boldsymbol{E}(\boldsymbol{r},t)\exp(\mathrm{i}\omega t)\mathrm{d}t \tag{5.12}$$

and the dielectric constant $\varepsilon(\omega)$ can be written as

$$\varepsilon(\omega) = 1 + \widetilde{\chi}^{(1)}(\omega) + \varepsilon_{NL} \tag{5.13}$$

$$\varepsilon_{NL} = \frac{3}{4}\widetilde{\chi}^{(3)}(\omega) \,|\, E(\boldsymbol{r},\omega)\,|^2 \tag{5.14}$$

According to Eq.(5.11), the theory of pulse propagation in nonlinear dispersive media in the slowly varying envelope approximation with the assumption that the spectral width of the pulse is much smaller than the frequency of the incident radiation can be obtained.

5.3.1 Mode Characteristics

At any frequency ω, optical fibers can support a finite number of guided modes whose spatial distributions $\widetilde{\boldsymbol{E}}(\boldsymbol{r},\omega)$ are a solution of the wave equation (5.11) and satisfy all appropriate boundary conditions. In addition, the fiber can support a continuum of unguided radiation modes. Although the inclusion of radiation modes is crucial in problems involving transfer of power between bounded and radiation modes, they do not play an important role in the discussion of nonlinear effects.

Because of the cylindrical symmetry of fibers, it is useful to express the wave equation in cylindrical coordinates. The wave equation for $\widetilde{E}_z$ is easily solved by using the method of separation of variables, resulting in the following general form

$$\widetilde{E}(r,\omega) = A(\omega)F(\rho)\exp(\pm \mathrm{i}m\phi)\exp(\mathrm{i}\beta z) \tag{5.15}$$

where A is a normalization constant, β is the propagation constant, m is an integer, and

$$F(\rho) = J_m(\kappa\rho), \quad \rho \leqslant a \tag{5.16}$$

$$F(\rho) = K_m(\gamma\rho), \quad \rho \geqslant a \tag{5.17}$$

In Eqs.(5.16) and (5.17), J_m is the Bessel function, K_m is the modified Bessel function, a is the core radius and

$$\kappa = (n_1^2 k_0^2 - \beta^2)^{1/2} \tag{5.18}$$

$$\gamma = (\beta^2 - n_2^2 k_0^2)^{1/2} \tag{5.19}$$

where n_1 and n_2 are the refractive index of fiber core and cladding layer, respectively.

The derivation of the eigenvalue equation whose solutions determine the propagation constant β for the fiber modes is left to the references, here we write the eigenvalue equation directly, i.e.,

$$\left[\frac{J_m'(\kappa a)}{\kappa J_m(\kappa a)} + \frac{K_m'(\gamma a)}{\gamma K_m(\gamma a)}\right]\left[\frac{J_m'(\kappa a)}{\kappa J_m(\kappa a)} + \frac{n_2^2}{n_1^2}\frac{K_m'(\gamma a)}{\gamma K_m(\gamma a)}\right] = \left[\frac{m\beta k_0(n_1^2 - n_2^2)}{a n_1 \kappa^2 \gamma^2}\right]^2 \tag{5.20}$$

where a prime denotes differentiation with respect to the argument.

The eigenvalue equation (5.20) in general has several solutions for β for each integer value of m. It is customary to express these solutions by β_{mn}, where both m and n take integer values. Each eigenvalue β_{mn} corresponds to one specific mode supported by the fiber. The corresponding modal field distribution is obtained from Eq.(5.15). It turns out that there are two types of fiber modes, designated as HE_{mn} and EH_{mn}. For m=0, these modes are analogous to the transverse-electric (TE) and transverse-magnetic (TM) modes of a planar waveguide because the axial component of the electric field, or the magnetic field, vanishes. However, for m>0, fiber modes become hybrid, i.e., all six components of the electromagnetic field are nonzero.

5.3.2 Optical Pulse Propagation and Pulse Spreading in Fibers

The study of most nonlinear effects in optical fibers involves the use of short pulses with widths ranging from ~10ns to 10fs. When such optical pulses propagate inside a fiber, both dispersive and nonlinear effects influence their shape and spectrum.

A basic equation that governs propagation of optical pulses in nonlinear dispersive fibers can be written as

$$\nabla^2 \boldsymbol{E} - \frac{1}{c^2}\frac{\partial^2 \boldsymbol{E}}{\partial t^2} = \mu_0 \frac{\partial^2 \boldsymbol{P}_L}{\partial t^2} + \mu_0 \frac{\partial^2 \boldsymbol{P}_{NL}}{\partial t^2} \tag{5.21}$$

It is necessary to make several simplifying assumptions before solving Eq.(5.21). First, $\boldsymbol{P}_{NL}$ is treated as a small perturbation to $\boldsymbol{P}_L$. This is justified because nonlinear changes in the refractive index are $<10^{-6}$ in practice. Second, the optical field is assumed to maintain its polarization along the fiber length so that a scalar approach is valid. This is not really the case, unless polarization-maintaining fibers are used, but the approximation works quite well in practice; Third, the optical field is assumed to be quasi-monochromatic, i.e., the pulse spectrum, centered at ω_0, is assumed to have a spectral width $\Delta\omega$ such that $\Delta\omega/\omega_0 \ll 1$. Since $\omega_0 \approx 10^{15}\text{s}^{-1}$, the last assumption is valid for pulses as short as 0.1ps.

Equation (5.11) can be solved by using the method of separation of variables. If we assume a solution of the form

$$\tilde{E}(\boldsymbol{r}, \omega - \omega_0) = F(x, y)\tilde{A}(z, \omega - \omega_0)\exp(\mathrm{i}\beta_0 z) \tag{5.22}$$

where $\tilde{A}$ is a slowly varying function of z and β_0 is the wave number. Substituting from Eq. (5.22) into Eq.(5.21), we obtain the following two equations

$$\frac{\partial^2 F}{\partial x^2} + \frac{\partial^2 F}{\partial y^2} + [\varepsilon(\omega)k_0^2 - \tilde{\beta}^2]F = 0 \tag{5.23}$$

$$2\mathrm{i}\beta_0 \frac{\partial \tilde{A}}{\partial z} + (\tilde{\beta}^2 - \beta_0^2)\tilde{A} = 0 \tag{5.24}$$

where

$$\varepsilon = (n + \Delta n)^2 \approx n^2 + 2n\Delta n \tag{5.25}$$

$$\tilde{\beta} = \beta(\omega) + \Delta\beta \tag{5.26}$$

$$\Delta n = n_2 |E|^2 + \frac{\mathrm{i}\tilde{\alpha}}{2k_0} \tag{5.27}$$

$$\Delta\beta = \frac{k_0 \iint_{-\infty}^{+\infty} \Delta n |F(x, y)|^2 \mathrm{d}x\mathrm{d}y}{\iint_{-\infty}^{+\infty} |F(x, y)|^2 \mathrm{d}x\mathrm{d}y} \tag{5.28}$$

Eq.(5.23) can be solved using first-order perturbation theory. We replace ε with n_2 and obtain the modal distribution $F(x, y)$, and the corresponding wave number $\beta(\omega)$. For a single-mode fiber, $F(x, y)$ corresponds to the modal distribution of the fundamental fiber mode HE_{11}, or by the Gaussian approximation. Equation (5.24) can further expressed as

$$\frac{\partial A}{\partial z}+\beta_1\frac{\partial A}{\partial t}+\frac{\mathrm{i}}{2}\beta_2\frac{\partial^2 A}{\partial t^2}+\frac{\alpha}{2}A=\mathrm{i}\gamma|A|^2 A \tag{5.29}$$

Equation (5.29) describes propagation of picosecond optical pulse in single-mode fibers. It is often referred to as the nonlinear Schrödinger (NLS) equation because it can be reduced to that form under certain conditions. It includes the effects of fiber losses through α, of chromatic dispersion through β_1 and β_2, and of fiber nonlinearity through γ.

The combined effects of group-velocity dispersion (GVD) and self-phase modulation (SPM) on optical pulses propagating inside a fiber can be studied by solving a pulse-propagation equation.

5.3.3 Dispersion Management

In a fiber-optic communication system, information is transmitted over a fiber by using a coded sequence of optical pulses whose width is determined by the bit rate B of the system. Dispersion-induced broadening of pulses is undesirable as it interferes with the detection process and leads to errors if the pulse spreads outside its allocated bit slot ($T_B=1/B$). Clearly, GVD limits the bit rate B for a fixed transmission distance L. The dispersion problem becomes quite serious when optical amplifiers are used to compensate for fiber losses because L can exceed thousands of kilometers for long-haul systems.

Even though operation at the zero-dispersion wavelength is most desirable from the standpoint of pulse broadening, other considerations may preclude such a design. For example, at most one channel can be located at the zero-dispersion wavelength in a wavelength-division-multiplexed (WDM) system. Moreover, strong four-wave mixing occurring when GVD is relatively low forces WDM systems to operate away from the zero-dispersion wavelength so that each channel has a finite value of β_2. Of course, GVD-induced pulse broadening then becomes of serious concern. The technique of dispersion management provides a solution to this dilemma. It consists of combining fibers with different characteristics such that the average GVD of the entire fiber link is quite low while the GVD of each fiber section is chosen to be large enough to make the four-wave-mixing effects negligible. In practice, a periodic dispersion map is used with a period equal to the amplifier spacing (typically 50~100km). Amplifiers compensate for accumulated fiber losses in each section. Between each pair of amplifiers, just two kinds of fibers, with opposite signs of β_2, are combined to reduce the average dispersion to a small value. When the average GVD is set to zero, dispersion is totally compensated.

When the bit rate of a single channel exceeds 100Gb/s, one must use ultrashort pulses (width~1ps) in each bit slot. For such short optical pulses, the pulse spectrum becomes broad enough that it is difficult to compensate GVD over the entire bandwidth of the pulse (because of the frequency dependence of β_2). The simplest solution to this problem is provided by fibers, or other devices, designed such that both β_2 and β_3 are compensated simultaneously. For a fiber link containing two different fibers of lengths L_1 and L_2, the conditions for broadband dispersion compensation are given by

$$\beta_{21}L_1 + \beta_{22}L_2 = 0 \quad \text{and} \quad \beta_{31}L_1 + \beta_{32}L_2 = 0 \tag{5.30}$$

where β_{2j} and β_{3j} are the GVD and TOD (third-order dispersion) parameters for fiber of length $L_j (j=1, 2)$.

It is generally difficult to satisfy both conditions simultaneously over a wide wavelength range. However, for a 1ps pulse, it is sufficient to satisfy Eq.(5.30) over a 4~5nm bandwidth. This requirement is easily met for dispersion-compensating fibers (DCFs), especially designed with negative values of β_3 (sometimes called reverse-dispersion fibers). Fiber gratings, liquid-crystal modulators, and other devices can also be used for this purpose.

5.3.4 Solitons

A fascinating manifestation of the fiber nonlinearity occurs through optical solitons, formed as a result of the interplay between the dispersive and nonlinear effects. The word "soliton" refers to special kinds of wave packets that can propagate undistorted over long distances. Solitons have been discovered in many branches of physics. In the context of optical fibers, not only are solitons of fundamental interest but they have also found practical applications in the field of fiber-optic communications. The use of solitons for optical communications was first suggested in 1973. By the year 1999, several field trials making use of fiber solitons have been completed.

The first-order soliton corresponds to the case of a single eigenvalue of nonlinear wave equations. It is referred to as the fundamental soliton because its shape does not change on propagation. Fundamental solitons can form in optical fibers at power levels available from semiconductor lasers even at a relatively high bit rate of 20Gb/s.

Starting in 1988, most of the experimental work on fiber solitons was devoted to their applications in fiber-optic communication systems. Such systems make use of fundamental solitons for representing "1" bits in a digital bit stream. In a practical situation, solitons can be subjected to many types of perturbations as they propagate inside an optical fiber. Examples of perturbations include fiber losses, amplifier noise (if amplifiers are used to compensate fiber losses), third-order dispersion, and intrapulse Raman scattering.

5.4 Impact of Fiber Nonlinearities

As seen in section 5.2.3, the nonlinear effects occurring inside optical fibers limit the maximum power levels and degrade the communication quality, including stimulated Brillouin scattering (SBS), stimulated Raman scattering (SRS), self-phase modulation (SPM) and cross-phase modulation (XPM), and four-wave mixing (FWM).

5.4.1 Stimulated Brillouin Scattering

SBS in optical fibers was first observed in 1972 and has been studied extensively since then because of its implications for lightwave systems. SBS generates a Stokes wave propagating in the backward direction. The frequency of the Stokes wave is downshifted by an amount that depends on the wavelength of incident signal. This shift is known as the Brillouin shift and is about 11 GHz in the wavelength region near 1.55μm. The intensity of the Stokes wave grows exponentially once the input power exceeds a threshold value.

Figure 5.5 shows variations in the transmitted reflected power (through SBS) for a 13-km-long dispersion-shifted fiber as the injected CW power is increased from 0.5 to 50 mW. No more than 3 mW could be transmitted through the fiber in this experiment after the onset of SBS.

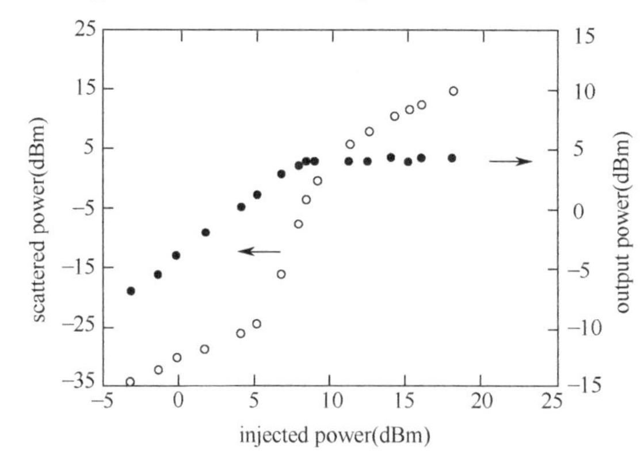

Fig.5.5 Output signal power (solid circles) and reflected SBS power (empty circles) as a function of power injected into a 13-km-long fiber

In lightwave systems, the optical signal is in the form of a time-dependent signal composed of an arbitrary sequence of 1 and 0 bits. One would expect the Brillouin threshold of such a signal to be higher than that of a CW beam. Considerable attention has been paid to estimating the Brillouin threshold and quantifying the SBS limitations for practical lightwave systems. The amount by which the threshold power increases depends on the modulation format used for data transmission. In the case of a coherent transmission scheme, the SBS threshold also depends on whether the amplitude, phase, or frequency of the optical carrier modulated for information coding. Most lightwave systems modulate amplitude of the optical carrier and use the so-called on–off keying scheme.

In modern wavelength-division-multiplexed (WDM) systems, fiber losses are compensated periodically using optical amplifiers. An important question is how amplifiers affect the SBS process. If the Stokes wave were amplified by amplifiers, it would accumulate over the entire link and grow enormously. Fortunately, periodically amplified lightwave systems typically employ an optical isolator within each amplifier that blocks the passage of the Stokes wave. However, the SBS growth between two amplifiers is still undesirable for two reasons. First, it removes power from the signal once the signal power exceeds the threshold level. Second, it induces large fluctuations in the remaining signal, resulting in degradation of both the SNR and the bit-error rate (BER). For these reasons, single-channel powers are invariably kept below the SBS threshold and are limited in practice to below 10 mW.

Fiber gratings can also be used to increase the SBS threshold. The Bragg grating is designed such that it is transparent to the forward-propagating pump beam, but the spectrum of the Stokes wave generated through SBS falls entirely within its stop band. A single grating, placed suitably in the middle, may be sufficient for relatively short fibers. Multiple gratings need to be used for long fibers. Another approach makes one try to minimize the overlap between the optical and acoustic modes through suitable dopants.

5.4.2 Stimulated Raman Scattering

SRS differs from SBS in several ways. First, it generates a forward-propagating Stokes wave. Second, the Raman shift by which the frequency of the Stokes wave is down-shifted is close to 13THz. Third, the Raman-gain spectrum is extremely broad and extends over a frequency range wider than 20THz. Fourth, the peak value of the Raman gain is lower by more than a factor of 100 compared with that of the Brillouin gain. SRS was first observed in optical fibers in 1972. Since then, the impact of SRS on the performance of lightwave systems has been studied extensively.

The Raman threshold, the power level at which the Raman process becomes stimulated and transfers most of the signal power to the Stokes wave, can be written as

$$P_{\text{th}} = 16A_{\text{eff}} / (g_{\text{R}} \cdot L_{\text{eff}}) \tag{5.31}$$

where we can replace L_{eff} with $1/\alpha$ for long fiber lengths used in lightwave systems. Using $g_{\text{R}}=1\times10^{13}$m/W, P_{th} is about 500 mW at wavelengths near 1.55μm. Since input powers are limited to below 10 mW because of SBS, SRS is not of concern for single-channel lightwave systems.

The situation is quite different for WDM systems, which transmit simultaneously multiple channels spaced 100GHz or so apart. The fiber link in this case acts as a Raman amplifier such that longer-wavelength channels are amplified by shorter-wavelength channels as long as their wavelength difference is within the Raman-gain bandwidth. The shortest-wavelength channel is depleted most as it can pump all other channels simultaneously. Such an energy transfer among channels can be detrimental for system performance as it depends on the bit pattern, occurring only when 1 bits are present simultaneously in the two channels acting as the pump and signal channels. The signal-dependent amplification leads to power fluctuations, which add to receiver noise and degrade the receiver performance.

The Raman crosstalk can be avoided if channel powers are made so small that Raman amplification is negligible over the fiber length. A simple model considered depletion of the highest-frequency channel in the worst case in which all channels transmit 1 bits simultaneously. A more accurate analysis should consider not only depletion of each channel but also its own amplification by shorter-wavelength channels. The effects of Raman crosstalk in a WDM system were quantified in a 1999 experiment by transmitting 32 channels, with 100-GHz spacing, over 100 km. At low input powers, SRS effects were relatively small and channel powers differed by only a few percent after 100 km. However, when the input power for each channel was increased to 3.6mW, the longest-wavelength channel had 70% more power than the shortest-wavelength channel. Moreover, the channel powers were distributed in an exponential fashion.

In long-haul lightwave systems, the crosstalk is also affected by the use of loss and dispersion-management schemes. Dispersion management permits high values of GVD locally while reducing it globally. Since the group-velocity mismatch among different channels is quite large in such systems, the Raman crosstalk should be reduced in a dispersion-managed system. In contrast, the use of optical amplifiers for loss management magnifies the impact of SRS-induced degradation. The reason is that inline amplifiers add broadband noise, which can seed the SRS process. As a result, noise is amplified along the link and results in degradation of the SNR. The SNR can be maintained if the channel power is increased as the number of amplifiers increases. The Raman-limited capacity of long-haul WDM systems depends on a large number of design parameters such as amplifier spacing, optical-filter bandwidth, bit rate, channel spacing, and total transmission distance.

Can Raman crosstalk be avoided by a proper system design? Clearly, reducing the channel power is the simplest solution, but it may not always be practical. An alternative scheme lets SRS occur over the whole link but cancels the Raman crosstalk by using the technique of spectral inversion. As the name suggests, if the spectrum of the WDM signal were inverted at some appropriate distance, short-wavelength channels would become long-wavelength channels and vice versa. As a result, the direction of Raman-induced power transfer will be reversed such that channel powers become nearly equal at the end of the fiber link. Complete cancelation of Raman crosstalk for a two-channel system requires spectral inversion at mid-span if GVD effects are negligible or compensated. The location of spectral inversion is not necessarily in the middle of the fiber span but changes depending on gain-loss variations. Spectral inversion can be accomplished through FWM inside a fiber to realize phase conjugation; the same technique is also useful for dispersion compensation.

5.4.3 Self-Phase Modulation

The intensity dependence of the refractive index leads to SPM-induced nonlinear phase shift, resulting in chirping and spectral broadening of optical pulses. Clearly, SPM can affect the performance of lightwave system. When SPM is included together with fiber dispersion and losses, the propagation of an optical bit stream through optical fibers is governed by the nonlinear Schrödinger (NLS) equation

$$\mathrm{i}\frac{\partial A}{\partial z}+\frac{\mathrm{i}\alpha}{2}A-\frac{\beta_2}{2}\frac{\partial^2 A}{\partial T^2}+\gamma|A|^2A=0 \tag{5.32}$$

where α, β_2 and γ govern the effects of loss, GVD and SPM, respectively. All three parameters become functions of z when loss- and dispersion-management schemes are employed for long-haul lightwave systems.

It is useful to eliminate the loss term in Eq.(5.32) with the transformation

$$A(z,T) = \sqrt{P_0}\,\mathrm{e}^{-\alpha z/2}U(z,T) \tag{5.33}$$

where P_0 is the peak power of input pulses. Eq.(5.32) then takes the form

$$\mathrm{i}\frac{\partial U}{\partial z} - \frac{\beta_2(z)}{2}\frac{\partial^2 U}{\partial T^2} + \gamma P_0 p(z)|U|^2 U = 0 \tag{5.34}$$

where power variations along a loss-managed fiber link are included through the periodic function $p(z)$. In the case of lumped amplifiers, $p(z)=\mathrm{e}^{\alpha z}$ between two amplifiers but equals 1 at the location of each amplifier. It is not easy to solve Eq.(5.34) analytically except in some simple cases. In the specific case of $p=1$and β_2 is constant but negative, this equation reduces to the standard NLS equation.

From a practical standpoint, the effect of SPM is to chirp the pulse and broaden its spectrum. The broadening factor can be estimated, without requiring a complete solution of Eq.(5.34), using various approximations. SPM enhances pulse broadening in the normal-GVD regime but leads to pulse compression in the anomalous-GVD regime. This behavior can be understood by noting that the SPM-induced chirp is positive in nature. As a result, the pulse goes through a contraction phase when $\beta_2<0$. This is the origin of the existence of solitons in the anomalous-GVD regime.

Another SPM-induced limitation results from the phenomenon of modulation instability occurring when the signal travels in the anomalous-GVD regime of the transmission fiber. At first sight, it may appear that modulation instability is not likely to occur for a signal in the form of a pulse train. In fact, it affects the performance of periodically amplified lightwave systems considerably. This can be understood by noting that optical pulses in a non-return-to-zero (NRZ) format system occupy the entire time slot and can be several bits long depending on the bit pattern. As a result, the situation is quasi-CW-like. As early as 1990, numerical simulations indicated that system performance of a 6000 km fiber link, operating at bit rates > 1Gb/s with 100 km amplifier spacing, would be severely affected by modulation instability if the signal propagates in the anomalous-GVD regime and is launched with peak power levels in excess of a few milliwatts.

SPM can lead to the degradation of SNR when optical amplifiers are used for loss compensation. Such amplifiers add to a signal broadband noise that extends over the entire bandwidth of amplifiers (or optical filters when they are used to reduce noise). Even close to the zero-dispersion wavelength, amplifier noise is enhanced considerably by SPM. In the case of anomalous GVD, spectral components of noise falling within the gain spectrum of modulation instability will be enhanced by this nonlinear process, resulting in further degradation of the SNR. Moreover, periodic power variations occurring in long-haul systems create a nonlinear index grating that can lead to modulation instability even in the normal-GVD regime. Both of these effects have been observed experimentally. In a 10Gb/s system, considerable degradation in system performance was noticed after a transmission distance of only 455km. In general, long-haul systems perform better when the average GVD of the fiber link is kept positive ($\beta_2>0$).

5.4.4 Cross-Phase Modulation

When two pulses of different wavelengths propagate simultaneously inside optical fibers, their optical phases are affected not only by SPM but also by XPM. The XPM effects are quite important for WDM lightwave systems because the phase of each optical channel depends on the bit patterns of all other channels.

The XPM effects occurring within a fiber amplifier are normally negligible because of a small length of doped fiber used. The situation changes for the L-band amplifiers, which operate in the 1570 to 1610nm wavelength region and require fiber lengths in excess of 100m. The effective mode area of doped fibers used in such amplifiers is relatively small, resulting in larger values of the nonlinear parameter γ and enhanced XPM-induced phase shifts. As a result, the XPM can lead to considerable power fluctuations within an L-band amplifier. A new feature is that such XPM effects are independent of the channel spacing and can occur over the entire bandwidth of the amplifier. The reason for this behavior is that all XPM effects occur before pulses walk off because of group-velocity mismatch.

The XPM effects can be reduced in modern WDM systems through the use of differential phase-shift keying (DPSK) format. The DPSK is often combined with the return-to-zero (RZ) format such that a pulse is present in every bit slot and the information is encoded only through phase variations. It is easy to understand qualitatively why XPM-induced penalties are reduced for such a lightwave system. The main reason why XPM leads to amplitude fluctuations and timing jitter when Amplitude Shift Keying format (ASK) format is used is related to the random power variations that mimic the bit pattern. It is easy to see that the XPM will be totally harmless if channel powers were constant in time because all XPM-induced phase shifts will be time-independent, producing no frequency and temporal shifts. Although this is not the case for an RZ-DPSK system, the XPM effects are considerably reduced because of a strictly periodic bature of the power variations. Physically speaking, all bits undergo nearly identical collision histories, especially if the average channel power does not vary too much along the link, resulting in negligible XPM-induced power penalties.

5.4.5 Four-Wave Mixing

Four-wave mixing (FWM) is a major source of nonlinear crosstalk for WDM lightwave systems. The physical origin of FWM-induced crosstalk, and the resulting system degradation, can be understood by noting that FWM can generate a new wave at the frequency $\omega_F=\omega_i+\omega_j+\omega_k$, whenever three waves of frequencies ω_i, ω_j, and ω_k copropagate inside the fiber. For an M-channel system, i, j, and k vary from 1 to M, resulting in a large combination of new frequencies generated by FWM. In the case of equally spaced channels, most new frequencies coincide with the existing channel frequencies and interfere coherently with the signals in those channels. This interference depends on the bit pattern and leads to considerable fluctuations in the detected signal at the receiver. When channels are not equally spaced, most FWM components fall in between the channels and add to overall noise. In both cases, system performance is affected by the loss in channel powers, but the degradation is much more severe for equally spaced channels because of the coherent nature of crosstalk.

In the case of equal channel spacing, most FWM components fall within the channel spectra and cannot be seen as clearly as in Fig.5.6 in the spectral domain.

A simple scheme for reducing the FWM-induced degradation consists of designing WDM systems with unequal channel spacings. The main impact of FWM in this case is to reduce the channel power. This power depletion results in a power penalty at the receiver whose magnitude can be controlled by varying the launched power and fiber dispersion. Experimental measurements on a WDM system, in which eight 10Gb/s channels were transmitted over 137km of dispersion-shifted fiber, confirm the advantage of unequal channel spacings. In a 1999 experiment, this technique was used to transmit 22 channels, each operating at 10Gb/s, over 320km of dispersion-shifted fiber with 80km amplifier spacing. Channel spacings ranged from 125 to 275GHz in the 1532 to 1562nm wavelength region and were determined using a periodic allocation scheme. The zero-dispersion wavelength of the fiber was close to 1548nm, resulting in near phase matching of many FWM components. Nonetheless, the system performed quite well (because of unequal channel spacings) with less than 1.5dB power penalty for all channels.

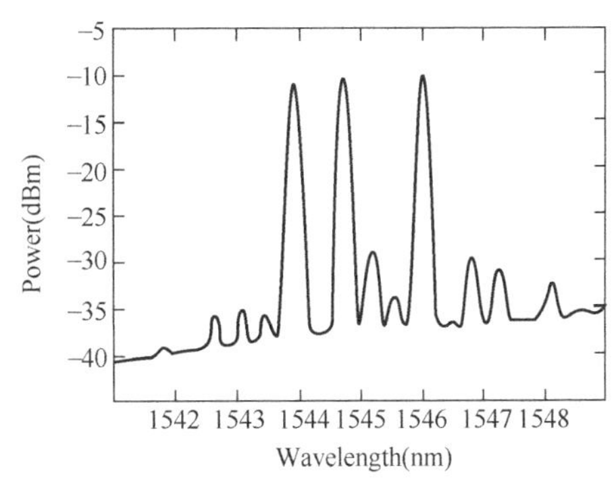

Fig.5.6 Optical spectrum measured at the output of a 25km long fiber when three channels, each with 3mW average power, are lunched into it

The use of nonuniform channel spacings is not always practical because many WDM components require equal channel spacings. Also, this scheme is spectrally inefficient since the bandwidth of the resulting WDM signal is considerably larger compared with the case of equally spaced channels. An alternative is offered by the dispersion-management technique discussed earlier. In this case, fibers with normal and anomalous GVD are combined to form a periodic dispersion map such that GVD is locally high all along the fiber even though its average value is quite low. As a result, the FWM efficiency is negligible throughout the fiber, resulting in little FWM-induced crosstalk. As early as 1993, eight channels at 10Gb/s could be transmitted over 280km by using dispersion management. By 1996, the use of dispersion management had become quite common for FWM suppression in WDM systems because of its practical simplicity. FWM can also be suppressed by using fibers whose GVD varies along the fiber length. The use of chirped pulses or carrier-phase locking is also helpful for this purpose.

5.5 All Optical Network

All Optical Networks are those in which either wavelength division or time division is used in new ways to form entire network structures where the messages travel in purely optical form all the way from one user location to another. There are many factors driving the development of high-capacity optical networks and their remarkably rapid transition from the research laboratories.

First and foremost is the continuing, relentless need for more capacity in the network. At the same time, businesses today rely on high-speed networks to conduct their businesses. These networks are used to interconnect multiple locations within a company as well as between companies for business-to-business transactions.

There is also a strong correlation between the increase in demand and the cost of bandwidth. Technological advances have succeeded in continuously reducing the cost of bandwidth. This reduced cost of bandwidth in turn spurs the development of a new set of applications that make use of more bandwidth and affects behavioral patterns.

Another factor causing major changes in the industry is the deregulation of the telephone industry. Traffic in networks is dominated by data as opposed to traditional voice traffic. In the past, the reverse was true, and so legacy networks were designed to efficiently support voice rather than data. Today, data transport services are pervasive and capable of producing quality of service to carry performances sensitive applications such as real-time voice and video.

Telecommunication networks have evolved during a century-long history of technological advances and social changes. Throughout the history, the digital network has evolved in three fundamental stages: asynchronous, synchronous, and optical.

The first digital networks were asynchronous networks, where each networks element's internal clock source timed its transmitting signal. Because each clock had a certain amount of variation, signals arriving and transmitting could have a large variation in timing, which often resulted in bit errors.

More importantly, as optical fiber deployment increased, no standards existed to mandate how network should format the optical signal. A myriad of proprietary methods appeared, making it difficult for network providers to interconnect equipment from different vendors.

The need for optical standards led to the creation of the synchronous optical network (SONET). SONET standardized line rates, coding schemes, bit-rates hierarchies, and operations and maintenance functionality. SONET also defined the types of network elements required, network architectures that vendors could implement, and the functionality that each node must perform. Network providers could now use different vendor's optical equipment with the confidence of at least basic interoperability.

The one aspect of SONET that has allowed it to survive during a time of tremendous changes in network capacity needs is its scalability. Base on its open-ended growth plan for higher bit rates, theoretically no upper limit exists for SONET bit rates. However, as higher bit rates are used, physical limitations in the laser sources and optical fiber begin to make the practice of endlessly increasing the bit rates on each signal an impractical solution. Additionally, connection to the networks through access rings has also had increased requirements. Customers are demanding more services and options, and are carrying more and different types of data traffic. To provide full end-to-end connectivity, a new paradigm was needed to meet all the high capacity and the varied needs.

5.5.1 Components

The components used in modern optical networks include couplers, lasers, photodetectors, optical amplifiers, optical switches, filters and multiplexers.

- **Coupler**

A coupler is a versatile device and has many applications in an optical network. The simplest directional coupler is used to combine and split signals in an optical network. A 2 × 2 coupler consists of two input ports and two output ports, as shown in Fig.5.7. The most commonly used couplers are made by fusing two fibers together in the middle—these are called fused fiber couplers. Couplers can also be fabricated using waveguides in integrated optics. A 2 × 2 coupler, shown in Figure5.7, takes a fraction α of the power from input 1 and places it on output 1 and the remaining fraction 1-α on output 2. Similarly, a fraction 1-α of the power from input 2 is distributed to output 1 and the remaining power to output 2. We call α the coupling ratio.

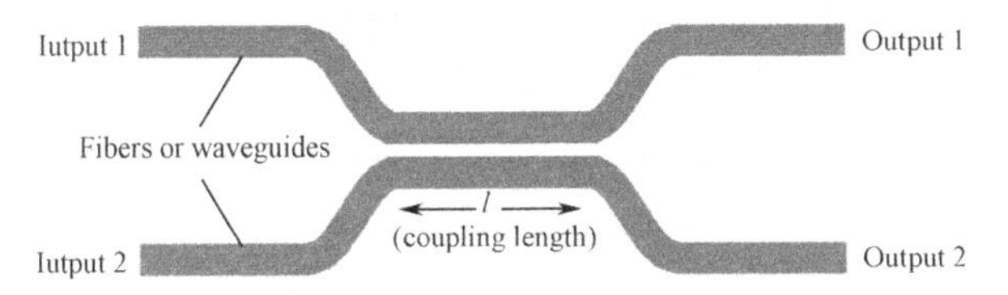

Fig.5.7 A directional coupler

Couplers are also used to trap off a small portion of the power from a light stream for monitoring purposes or other reasons. It can be designed to be either wavelength selective or wavelength independent over a usefully wide range.

- **Photodetectors**

The basic principle of photodetection is illustrated in Fig.5.8. Photodetectors are made of semiconductor materials. Photons incident on a semiconductor are absorbed by electrons in the valence band. As a result, these electrons acquire higher energy and are excited into the conduction band, leaving behind a hole in the valence band. When an external voltage is applied to the semiconductor, these electron-hole pairs give rise to a photocurrent. Photodetectors generate an electrical current proportional to the incident optical power.

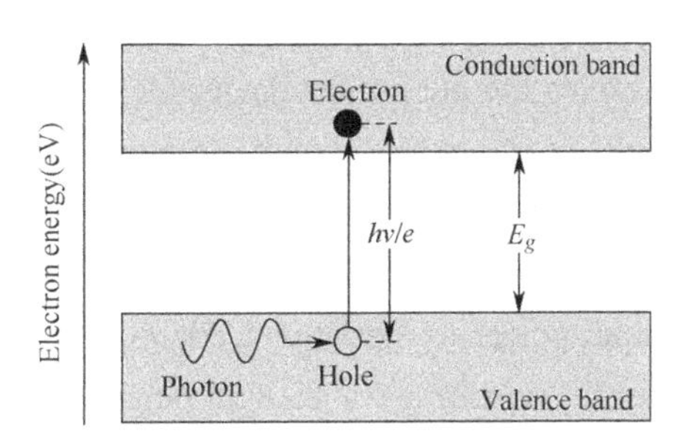

Fig.5.8 The basic principle of photodetectors using a semiconductor

The fraction of the energy of the optical signal that is absorbed and gives rise to a photocurrent is called the efficiency η of the photodetector. For transmission at high bit rates over long distances, optical energy is scarce, and thus it is important to design the photodetector to achieve an efficiency η of as close to 1 as possible. This can be achieved by using a semiconductor slab of sufficient thickness. The power absorbed by a semiconductor slab of thickness L can be written as

$$P_{abs} = (1 - \mathrm{e}^{-\alpha L})P_{in} \tag{5.35}$$

where P_{in} is the incident optical signal power, and α is the absorption coefficient of the material.

Then, we can readily obtain the efficiency of the photodetector

$$\eta = \frac{P_{abs}}{P_{in}} = (1 - \mathrm{e}^{-\alpha L}) \tag{5.36}$$

The absorption coefficient depends on the wavelength. However, photodetectors have a very wide operating bandwidth since a photodetector at some wavelength can also serve as a photodetector at all smaller wavelength.

● **Optical amplifiers**

In an optical communication system, the optical signals from the transmitter are attenuated by the optical fiber as they propagate through it. Other optical components such as multiplexers and couplers, also add loss. After some distance, the cumulative loss of signal strength causes the signal to become too weak to be detected. Before this happens, the signal strength has to be restored. Prior to the advent of optical amplifiers over the last decade, the only option was to regenerate the signal, that is, receive the signal and retransmit it. This process is accomplished by regenerators. A regenerator converts the optical signal to an electrical signal, cleans it up, and converts it back into an optical signal for onward transmission.

Optical amplifiers offer several advantages over regenerators. On one hand, regenerators are specific to the bit rate and modulation format used by the communication system. On the other hand, optical amplifiers are insensitive to the bit rate or signal formats. Thus a system using optical amplifiers can be more easily upgraded, for example, to a higher bit rate, without replacing the amplifiers. In contrast, in a system using regenerators, such an upgrade would require all the regenerators to be replaced. Furthermore, optical amplifiers have fairly large gain bandwidths, and as a consequence, a single amplifier can simultaneously amplify several WDM signals.

In contrast, we would need a regenerator for each wavelength. Thus optical amplifiers have become essential components in high-performance optical communication systems. Amplifiers, however, are not perfect devices. They introduce additional noise, and this noise accumulates as the signal passes through multiple amplifiers along its path due to the analog nature of the amplifier. The spectral shape of the gain, the output power, and the transient behavior of the amplifier are also important considerations for system applications. Ideally, we would like to have a sufficiently high output power to meet the needs of the network application. We would also like the gain to be flat over the operating wavelength range and to be insensitive to variations in input power of the signal.

● **Switches**

Optical switches are used in optical networks for a variety of applications. One application of optical switches is in the provisioning of lightpaths, used inside wavelength crossconnects to reconfigure them to support new lightpaths. Another important application is that of protection switching. Here the switches are used to switch the traffic stream from a primary fiber onto another fiber in case the primary fiber fails. The entire operation must typically be completed in several tens of milliseconds, which includes the time to detect the failure, communicate the failure to the appropriate network elements handling the switching, and the actual switch time. Thus the switching time required is on the order of a few milliseconds. In these networks, switches are used to switch signals on a packet-by-packet basis.

Switches are also important components in high-speed optical packet-switched networks. In these networks, switches are used to switch signals on a packet-by-packet basis. For this application, the switching time must be much smaller than the packet duration, and large switches will be needed. For example, ordinary Ethernet packets have lengths between about 60 to 1500 bytes. At

10Gb/s, the transmission time of a 60 byte packet is 48 ns. Thus, the switching time required for efficient operation is on the order of a few nanoseconds.

What's more, the different applications require different switching times and number of switch ports, as summarized in Table 5.1.

Yet another use for switches is as external modulators to turn on and off the data in front of a laser source. In this case, the switching time must be a small fraction of the bit duration. So an external modulator for a 10Gb/s signal (with a bit duration of 100ps) must have a switching time (or, equivalently, a rise and fall time) of about 10ps.

Table 5.1 Applications for optical switches and their switching time and port count requirements

Application	Switching Time Required	Number of ports
Provisioning	1~10 ms	>1000
Protection switching	1~10 ms	2~1000
Packet switching	1 ns	>100
External modulation	10 ps	1

5.5.2 Modulations and Demodulations

Our goal in this section is to introduce modulations and demodulations of digital signals.

● **Modulations**

Modulation is the process of converting data in electronic form to optical form for transmission on the fiber.

It can be divided into two categories: direct modulation and external modulation. Direct modulation of the laser or light emitting diode (LED) source can be used for transmission at low bit rates over short distances, whereas external modulation is needed for transmission at high bit rates over long distances.

The most commonly used modulation scheme in optical communication is on-off keying (OOK), which is illustrated in Fig.5.9.

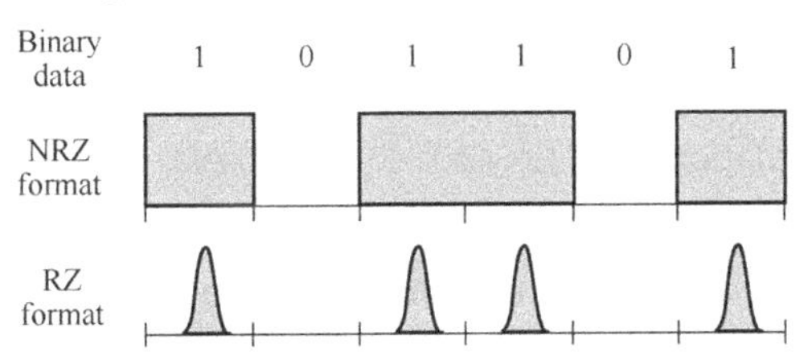

Fig.5.9 On-off keying modulation of binary digital data

In this modulation scheme, a 1 bit is encoded by the presence of a light pulse in the bit interval or by turning a light source (laser or LED) "on". A 0 bit is encoded (ideally) by the absence of a light pulse in the bit interval or by turning a light source "off". The bit interval is the interval of time available for the transmission of a single bit.

The optical signal emitted by a laser operating in the 1310 or 1550 nm wavelength band has a center frequency around 10^{14} Hz. This frequency is the optical carrier frequency. In what we have studied so far, the data modulates this optical carrier. In other words, with an OOK signal, the optical carrier is simply turned on or off, depending on the bit to be transmitted. Instead of modulating the optical carrier directly, we can have the data first modulate an electrical carrier in the microwave frequency range, typically ranging from 10MHz to 10GHz, as shown in Fig.5.10. The upper limit on the carrier frequency is determined by the modulation bandwidth available from the transmitter. The modulated microwave carrier then modulates the optical transmitter. If the transmitter is directly modulated, then changes in the microwave carrier amplitude get reflected as changes in the transmitted optical power envelope, as shown in Fig.5.10.

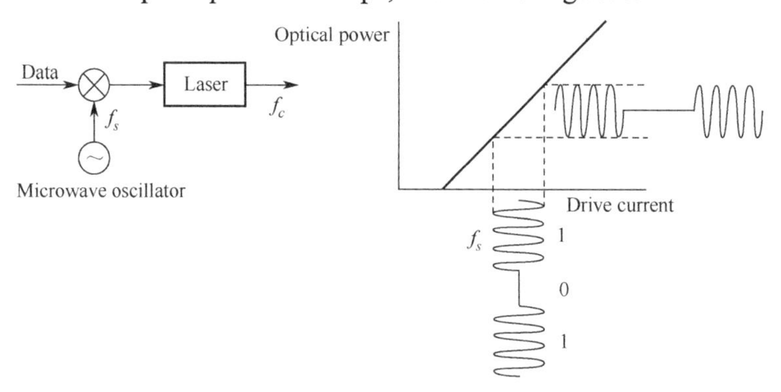

Fig.5.10 Subcarrier modulation. The data stream first modulates a microwave carrier, which, in turn, modulates the optical carrier.

The microwave carrier can itself be modulated in many different ways, including amplitude, phase, and frequency modulation, and both digital and analog modulation techniques can be employed. The figure shows an example where the microwave carrier is amplitude modulated by a binary digital data signal. The microwave carrier is called the subcarrier, with the optical carrier being considered the main carrier. This form of modulation is called subcarrier modulation.

- **Demodulations**

Demodulation is inverse process of modulation, i.e., converting optical signals back into electronic form and extracting the data that is transmitted. The modulated signals are transmitted over the optical fiber where they undergo attenuation and dispersion, have noise added to them from optical amplifiers, and sustain a variety of other impairments.

In principle, the demodulation process can be quite simple. Ideally, it can be viewed as"photon counting". In practice, there are various impairments that are not accounted for by this model. The receiver looks for the presence or absence of light during a bit interval. If no light is seen, it infers that a 0 bit was transmitted, and if any light is seen, it infers that a 1 bit was transmitted. This is called direct detection. Unfortunately, even in the absence of other forms of noise, this will not lead to an ideal error-free system because of the random nature of photon arrivals at the receiver. In practice, most receivers are not ideal, and their performance is not as good as that of the ideal receiver because they must contend with various other forms of noise, as we shall soon see.

The main complication in recovering the transmitted bit is that in addition to the photocurrent due to the signal there are usually three other additional noise currents. The first is the thermal noise current due to the random motion of electrons that is always present at any finite temperature. The second is the shot noise current due to the random distribution of the electrons generated by the photodetection process even when the input light intensity is constant. The shot noise current, unlike the thermal noise current, is not added to the generated photocurrent but is merely a convenient representation of the variability in the generated photocurrent as a separate component. The third source of noise is the spontaneous emission due to optical amplifiers that may be used between the source and the photodetector.

- **Noise**

The most common noises are the front-end amplifier noise and the avalanche photodiodes noise.

We saw that the photodetector is followed by a front-end amplifier. Components within the front-end amplifier, such as the transistor, also contribute to the thermal noise. This noise contribution is usually stated by giving the noise figure of the front-end amplifier. The noise figure Fn is the ratio of the input signal-to-noise ratio (SNRi) to the output signal-to-noise ratio (SNRo). Equivalently, the noise figure F_n of a front-end amplifier specifies the factor by which the thermal noise present at the input of the amplifier is enhanced at its output. Thus the thermal noise contribution of the receiver has variance thermal

$$\sigma_{thermal}^2=\frac{4k_BT}{R_L}F_nB_e \tag{5.37}$$

when the front-end amplifier noise contribution is included. Typical values of F_n are 3~5 dB.

The avalanche gain process in avalanche photodiodes has the effect of increasing the noise current at its output. This increased noise contribution arises from the random nature of the avalanche multiplicative gain, $G_m(t)$. This noise contribution is modeled as an increase in the shot noise component at the output of the photodetector. If we denote the responsivity of the APD by APD, and the average avalanche multiplication gain by G_m, the average photocurrent is given by $\bar{I}=R_{APD}P=G_mRP$, and the shot noise current at the APD output has variance

$$\sigma_{shot}^2=2eG_m^2F_A(G_m)RPB_e \tag{5.38}$$

The quantity $F_A(G_m)$ is called the excess noise factor of the APD and is an increasing function of the gain G_m.It is given by

$$F_A(G_m)=k_AG_m+(1-k_A)(2-1/G_m) \tag{5.39}$$

The quantity k_A is called the ionization coefficient ratio and is a property of the semiconductor material used to make up the APD. It takes values in the range 0~1. The excess noise factor is an increasing function of k_A, and thus it is desirable to keep k_A small. The value of k_A for silicon (which is used at 0.8μm wavelength) is 1, and for InGaAs (which is used at 1.3 and 1.55μm wavelength bands) it is 0.7. Note that F_A (1)=1, and thus Eq.(5.38) also yields the shot noise variance for a pin receiver if we set G_m=1.

- **Capacity limits of optical fiber**

An upper limit on the spectral efficiency and the channel capacity is given by Shannon's theorem. Shannon's theorem says that the channel capacity C for a binary linear channel with additive noise is given by

$$C = B\log_2(1 + S / N) \tag{5.40}$$

Here B is the available bandwidth and S/N is the signal-to-noise ratio. A typical value of S/N is 100. Using this number yields a channel capacity of 350 Tb/s or an equivalent spectral efficiency of 7b/s/Hz. Clearly, such efficiencies can only be achieved through the use of multilevel modulation schemes.

In practice, today's long-haul systems operate at high power levels to overcome fiber losses and noise introduced by optical amplifiers. At these power levels, nonlinear effects come into play. These nonlinear effects can be thought of as adding additional noise, which increases as the transmitted power is increased. Therefore they in turn impose additional limits on channel capacity. Recent work to quantify the spectral efficiency, taking into account mostly cross-phase modulation, shows that the achievable efficiencies are of the order of 3~5b/s/Hz. Other nonlinearities such as four-wave mixing and Raman scattering may place further limitations. At the same time, we are seeing techniques to reduce the effects of these nonlinearities. Another way to increase the channel capacity is by reducing the noise level in the system. The noise figure in today's amplifiers is limited primarily by random spontaneous emission, and these are already close to theoretically achievable limits. Advances in quantum mechanics may ultimately succeed in reducing these noise limits.

5.5.3 WDM Network Design

There is a clear benefit to build wavelength-routing networks, as opposed to simple point-to-point WDM links. The main benefit is that traffic that is not to be terminated within a node can be passed through by the node, resulting the significant saving in high-layer terminating equipment.

The WDM network provides circuit-switched lightpaths that can have varying degrees of transparency associated with them. Wavelengths can be reused in the network to support multiple lightpaths as long as no two lightpaths are assigned the same wavelength on a given link. Lightpaths may be protected by the network in the event of failures. They can be used to provide flexible interconnections between users of the optical network, such as IP routers, allowing the router topology to be tailored to the needs of the router network.

An optical line terminal (OLT) multiplexes and demultiplexes wavelengths and is used for point-to-point applications. It typically includes transponders, multiplexers, and optical amplifiers. Transponders provide the adaptation of user signals into the optical layer. They also constitute a significant portion of the cost and footprint in an OLT. In some cases, transponders can be eliminated by deploying interfaces that provide already-adapted signals at the appropriate wavelengths in other equipment.

The design of optical network is more complicated than the design of traditional networks. It includes the design of the higher-layer topology (IP or SONET), which is the lightpath topology design problem, and its realization in the optical layer, which is the routing and wavelength assignment problem. These problems may need to be solved in conjunction if the carrier provides IP or SONET VTs over its own optical infrastructure. However, this is difficult to do, and a practical approach may be to iteratively solve these problems.

We must concern attention to the wavelength dimensioning problem. The problem here is to provide sufficient capacity on the links of the wavelength-routing network to handle the expected demand for lightpaths. This problem is solved today by periodically forecasting a traffic matrix and (re)designing the network to support the forecasted matrix. Alternatively, you can employ statistical traffic demand models to estimate the required capacities, and we discussed two such models.

Words and Expressions

access	使用权
accumulate	累积
acoustic	听觉的，声音的，原声的
advent	出现
allocation	分配
analogous	相似的
annihilate	消灭
anomalous	反常的
applicable	可应用（实施）的；适当的，合适的
arbitrary	任意的
asynchronous	异步的
attenuation	衰减
avalanche	雪崩
bandwidth	带宽
bature	本质
bending loss	弯曲损耗
bound electron	束缚电子
boundary loss	边界损耗
category	种类，类，类别
cease	停止，中止
chromatic dispersion	色散
cladding layer	包层
coefficient	系数
coincide	一致，符合
compensate	补偿
confine	约束，限制
constituent	组成，构成
contend	竞争
core	纤芯
coupler	连接者，耦合器
cross-phase modulation	交叉相位调制
crosstalk	串扰

curve	曲线，弧线；弯曲部
cylindrical coordinate	柱坐标系
depletion	消耗，损耗
deploy	配置，部署
deregulation	违反规定，反常
derivative	派（衍）生物
detrimental	有害的
diagram	图解，简图，图表
dielectric	电介质，绝缘体
dilemma	困境
dispersion	分散；传播；消散；色散
dominantly	主导地，占优势地
electron-hole pair	空穴-电子对
enormously	巨大地，庞大地
enhance	加强，增大
envelope	包络
erbium-doped fiber amplifier	掺铒光纤放大器
essentially	实质上
evolve	发展，演变
exponentially	指数地
extensively	广泛地
fluctuations	涨落，波动
Foremost	首要的，第一的
govern	统治；管理；支配；影响
graded-index fiber	梯度折射率光纤
grating	光栅
group-velocity mismatch	群速度失配
haul	拖曳，拖
harmonic	调和的；调波
heterodyne	外差
heterodyne detector	外差检波器
hierarchies	分层
homodyne	零差（拍）
homodyne detector	零拍检波器
hybrid	混合的，复合的
impair	损害，损伤，削弱
implication	牵连，关联
impose	施加
infrastructure	结构，基础设施
insensitive	麻木的；冷淡的

interoperability	互操作性，互用性
inversion	反演
invert	倒转
intrapulse	脉冲间隔内的
isolator	隔离器
legacy	遗产、遗赠
long-haul	长途的，长距离的
mandate	授权
manifest	显现，表明，表现
matrix	模子，矩阵
mimic	模仿，模拟
modal dispersion	模间色散
moderate	适度的
modify	修改，修正
modulation	调制
multiplexer	多路器
myriad	无数的
NLS equation	非线性薛定谔方程
nonetheless	尽管如此
normalized frequency	归一化频率
novel	新颖的
onset	开始，起始
optical fiber	光纤
oscillation	振荡
packet	波包
paradigm	示例
pattern	模式，图案
perturbation	烦乱，扰乱
pervasive	普遍的，到处的
photocurrent	光电流
power penalty	功率损失，功率代价
profile	剖面，包络
proprietary	私有的
provision	规定，准备
quantum	定量
refer to as	称为，称作
regenerate	重建；再生
regime	制度
relative core-cladding index difference	相对纤芯-包层折射率差

relentless	不间断的
repeater	中继器
repeater spacing	中继距离
resonance	共振
restrict	限制，约束
reverse	反面的
reyleigh scattering	瑞利散射
route	途径
router	路由器
scalability	可扩展性
scarce	缺乏的
schematically	图示的
scheme	计划，方案
sequence	次序，顺序
silica（SO_2 silium dioxide）	二氧化硅
simultaneously	同时，一起
single-mode	单模
slab	厚板，平板，厚片
soliton	（光）孤子
spacings	间距，空隙
spectral	光谱的
splice loss	连接损耗
spontaneous	自发的
spontaneous emission	自发辐射
spurs	激励
step-index fiber	阶跃折射率光纤
susceptibility	极化率
sustain	维持，支撑
symmetry	对称（性）
synchronous	同步的
tailor	裁剪，使合适
thermal	热的，由热造成的
topology	拓扑学
transaction	交易，事务
transparent	透明的
transponders	转发器，应答器
tremendous	极大的，巨大的
trial	试验
tunable semiconductor laser	可调谐的半导体激光器
ultrashort	超短波的

valence　　某物结合、反映或作用于另一物的能力
variance　　矛盾，不同
vendors　　供应商
versatile　　通用的，用途广泛的
vibration　　振动
walk-off　　离散，走离

Grammar　专业英语翻译方法（五）：名词的译法

英语和汉语一样，都有“一词多义”和“一义多词”的特点。

例：light

—day light　　日光
—light weight　　轻的重量
—to light the lamp　　点燃这盏灯

名词的分类译法

1．意译法

guided missile　　导弹
holography　　全息照相术
jet plane　　喷气式飞机
loudspeaker　　扬声器
monocrystal　　单晶（体）
radioactive isotope　　放射性同位素

2．音译法

bit　　比特
hertz　　赫兹
maxwell　　麦克斯韦

3．直译法

spend money like water　　花钱如流水
chain reaction　　连锁反应

4．音意兼译法

card　　卡片
logic　　逻辑
tank　　坦克
microfarad　　微法（电容单位）

由人名组成的术语：

Brinell hardness　　布氏硬度

Curie point　　居里点
Doppler shift　　多普勒频移
Geiger counter　　盖格计数器
Kelvin bridge　　凯尔文电桥
Lenz's law　　楞次定律
Morse code　　莫尔斯电码

5．形译法

cross wire　　十字线
T-square　　丁字尺
stepped waveguide　　阶梯形波导
twist drill　　麻花钻
V-belt　　三角皮带
zigzag wave　　锯齿形波

6．合成词

maser-microwave amplifier with stimulate emission of radiation　　受激辐射式微波放大器
radar-radio detecting and ranging　　雷达（无线电探测及测距）
laser-light amplification by stimulated emission of radiation　　激光器（受激辐射式光频放大器）

Part 6 Optical Components

This chapter focuses on some optical components commonly used in optical systems, such as optical media, spectral filters, integrated optical components, photoelectric detectors and thermal detectors etc..

6.1 Optical Media

6.1.1 Optical Glasses

Optical glasses are the most popular medium in optical systems. They are relatively easy to manufacture and to work; they can be made with highly desirable optical and mechanical characteristics and many of them are very stable under ordinary conditions. They can be manufactured in sizes up to a meter in diameter (in some cases to almost four meters) for lenses and considerably larger of mirrors.

These glasses may be manufactured in a continuous process, yielding fair refractive index uniformity at economical prices. However, the classical method of producing the glass in crucibles, a batch at a time, is still necessary for the highest index uniformity.

Glass is defined as an inorganic product of fusion cooled to a rigid condition without crystallizing. It is obtained from a melt that is cooled in such a way that the molecular structure is irregular over long ranges, although it may be quite regular over a number of interatomic spaces. For instance, when crystalline quartz is melted and cooled slowly, a silica glass, "fused silica", is obtained. Its chemical formula is SiO_2 and its molecular structure is tetrahedral (a silicon atom with four oxygen atoms forming a regular tetrahedron around it). These tetrahedral are joined at their corners, so that each oxygen atom is share by two tetrahedral; these, in turn, forming a larger interconnected network. Whereas in a crystal the relative angular orientation of neighboring tetrahedral is fixed, in a glass it varies from joint to joint.

6.1.2 Crystalline Optical Media

As optical media, crystals are much less popular than glasses for the obvious reason that they are far more difficult to manufacture in large size. Their use is invariably motivated by the need for characteristics not readily obtainable in glass. Among these characteristics are the following: (1) Spectral transmittance characteristics extending beyond those obtainable with the usual glasses; (2) Special shape of the dispersion curve; (3) Birefringence, electro-, and magnetooptical characteristics; (4) Mechanical and thermal strength.

The earliest isotropic crystalline material used in optical design is probably fluorite (calcium fluoride, CaF_2), which is outstanding both in spectral range (0.13~12μm) and in its dispersion characteristics, which, in combination with optical glasses, permit elimination of the secondary spectrum and apochromatization. Quartz, too, is a relatively popular material, combining a broad transmittance spectrum with great mechanical strength. It is both birefringent and optically active.

6.1.3 Optical Plastics

As the manufacturing technology of plastic materials advanced sufficiently to permit good stability and casting precision, these materials became increasingly significant as economical optical media. Their major advantages are as follows: (1) Manufacturing economy: Precision casting can replace grinding and polishing. Mounting devices can be manufactured integrally with the optical component. (2) Optical Advantages: High relative dispersion is feasible even at relatively low refractive indices; many plastics are superior to silicate glasses in UV and IR transmission. (3) Mechanical advantages: The greater impact strength and lighter weight of plastics make them superior to glass in some applications; the specific gravity of plastics is generally about half that of glass.

On the other hand, there are several important disadvantages: (1) Lower precision: The economical forming method, casting, limits the precision. (2) Mechanical limitations: Plastics are far less resistant to scratching and other forms of surface damage. (3) Thermal effects: Thermal effects in plastics are far more pronounced than they are in silicate glasses, both in relative expansion and in refractive index variation. (4) Limited range of refractive indices: Optical plastics with refractive index above 1.6 do not seem to be practical.

6.2 Spectral Filters

Spectral filters are devices used to modify the (spectral) transmittance of an optical system. They may be made of a homogeneous material with appropriate spectral absorption characteristics, or use other optical phenomenon, such as multiple-beam interference or optical anisotropy, to effect the modification.

6.2.1 Filtering Characteristics

The optical properties of filters may be divided into integral and spectral characteristics. The spectral characteristics again fall into several subcategories: (1) Sharp-cutting (high-pass or low-pass); (2) Band-pass (wide or narrow); (3) Compensating; (4) Neutral density.

In integral terms the performance of a filter may be specified in terms of chromaticity and total transmittance.

Sharp-cutting transmission filters pass flux, more or less completely, below a certain wavelength, and absorb (or reflect) the flux above this wavelength, or vice versa. Accordingly, such filters are classified as short wavelength pass or long wavelength pass (short-pass and long-pass), respectively. For filters used in reflection, this terminology may become confusing and should be made specific (e.g., "long-reflecting", etc.). Filters whose spectral characteristics change from transmitting to reflecting at a certain wavelength are called dichroic, that is, they are short-pass in transmittance and long-pass in reflection, or vice versa. In addition to the class of the sharp-cutting filter, the spectral position of the transition and the rate of transition are usually required.

Of special interest are short-pass filters with the cutoff point just inside the IR. They are essentially clear in the visible but absorb heavily throughout the IR and serve as heat absorbing filters.

The characteristics of band-pass filters are often defined in terms of the half-peak transmittance points, that is, the wavelength values at which the transmittance has dropped to half its peak value. The midpoint between these defines the position of the passband and the spectral range between them, the bandwidth. To define the sharpness of the passband characteristics, the range between the 5% transmittance points, too, is sometimes given. The rejection range, occasionally called the free filter range, is the spectral range, outside the base bandwidth range, over which the transmittance remains below a certain value, here taken as 1%. The filter efficiency may be defined in terms of the peak transmittance value.

Band-rejection filters are transparent over a wide spectral region and absorb radiation in a narrow range inside this region. They are defined in terms of the rejection band in a manner analogous to the definition of band-pass filters. Here the range of rejection replaces the transparency range in specifying their utility range.

Filters having slowly changing transmittance spectra are generally classified as compensating filters. Their function is usually to modify somewhat the spectral luminance of a light source or sensitivity spectrum of a detector.

Neutral density filters may be specified in terms of their mean density, their range of neutrality, and the magnitude of the transmittance variation over this range. Note that, in addition to the usual dye filters (glass or gelatine), partially exposed and developed silver halide emulsion, evaporated metal film, crossed polaroid filters, and fine metal screens are occasionally convenient neutral-density filters.

6.2.2 Absorption Filters

Absorption filters are made by dissolving in glass dyes or metal ions that cause light to be absorbed over certain spectral regions. Dyes may also be dissolved in gelatine, which is subsequently formed into a thin sheet to act as a filter. When the absorption is uniform over that visible region of the spectrum, the filter appears neutral; if it varies over this region, it appears cooled, in general.

The absorption may be due to one of three causes: (1) For substances dissolved in the glass: absorption due to interatomic or molecular bonds. (2) For substances forming colloidal suspensions: scattering and subsequent absorption by the suspended particles. (3) For substances forming microscopic suspensions: absorption by the suspended crystals.

Heating the glass tends to strengthen the absorption bands and to shift them toward the longer wavelength region: the loosened interatomic bonds correspond to lower resonant frequencies and, therefore, longer wavelengths. Subsequent quenching may freeze such a state into the temperatures due to a shifting of the oxidation-reduction equilibria and changed electron mobility.

Irradiation can change the state of the color centers (sites of imperfections in the atomic network) in the glass and hence its absorption spectrum.

6.2.3 Thin-Film Filters

The popular simple interference filter is essentially a Fabry-Perot etalon, where the spacer is a thin layer of dielectric material deposited by evaporation between two layers of partially reflecting metallic coating. But, whereas in the Fabry-Perot interferometer the two surfaces must be extremely flat (deviations of $\lambda/40$ accumulate to half a wavelength after 10 reflections), the requirements are far lower in the interference filter; here the evaporated layers follow the surface contour, so that the two surfaces are essentially parallel, tending to compensate mutually for any deviation from planeness.

An interference filter whose reflecting layers each have a reflectance and transmittance of 90% and 6%, respectively, will yield a peak transmittance of 36% and a relative bandwidth of 1.7% when used in second order (3.4%, when used in the first order). Its usable field extends ±7.5°; when it is used near an image plane, the equivalent numerical aperture is 0.13. These values are quite representative of commercially available interference filters.

By increasing the order, at any given wavelength, the passband becomes narrower, but the undesirable adjacent passbands come closer.

Higher efficiency, combined with higher rejection ratio can be obtained by substituting multiple dielectric layers for the metallic reflection layers. The dielectric reflecting layers exhibit very low absorption and therefore facilitate higher reflectance values at a given transmittance level. A representative filter of this type has peak transmittance of 80% and a relative bandwidth of 1.25% in the first order.

The periodic multiplayer film is relatively easy to analyze and yet permits considerable flexibility in synthesizing band-pass and sharp-cutting filters.

In general, a periodic multiplayer film with a two-component period has a reflectance peak at the wavelength for which the total period is optically half of a wavelength thick; the spectral width of the high-reflectance region increases with the ratio of the refractive indices of the two components. Thus, basically, such a multiplayer film leads itself to block a spectral band in transmission. A typical application might be a safety filter for use with a laser operating at a fixed wavelength. Here we would like to have high transmission throughout the visible part of the spectrum, with only a narrow region, surrounding the reflection peak, blocked. This can be accomplished by a simple two-component periodic stack having a small reflective index difference between the two components. Simultaneously, the same filter can be considered as a band-pass filter when it is used as a mirror in an optical system.

It is frequently desirable to split the flux passing an optical system into long- and short-wavelength portions, without absorbing either portion. This can be accomplished by dielectric multiplayer mirrors that operate as long- or short-wavelength pass filters, with both the transmitted and reflected portions of the flux continuing along their separate paths after reflection has taken place. Such mirrors are called dichroic.

Some applications of dichroic mirrors include: (1) In the projection of photographic slides and motion picture films, it is important to illuminate the transparency to be projected with the maximum luminous flux possible and to avoid illumination by flux outside the visible spectrum, primarily the IR flux, which only heats the transparency, possibly dangerously, without contributing to the visibility of the projected image. Often, in such systems, spherical (or ellipsoidal) mirrors are used to reflect an image of the source back onto itself and into the entrance pupil of the projection lens. It is then desirable to have the mirror reflect only the visible portion of the flux and to transmit the IR. This is preferable to having the mirror absorb the IR, being heated thereby, since the removal of heat from the immediate vicinity of the lamp is often a major problem in optical projection system design. Such a mirror is referred to as a “cold” mirror. For additional projection, an ir-reflecting, or “hot” mirror may be interposed between the light source and the transparency. Here, again, an IR-reflecting window is preferable to the more common ir-absorbing one for the reasons stated. (2) In colorimetry, it may be desired to split the flux into three spectral bands to measure their relative flux content by means of three different detectors. This may be accomplished by means of two long- or short-pass filters, or a combination of these. (3) As an illustration of a more sophisticated task, we cite a radar display where it may be desired to view a phosphor display screen and, simultaneously, to photograph it. The phosphor should have a high radiance-time product in the part of the spectrum where the photographic film is sensitive (blue and UV) and should have a long persistence in the region of great visual efficacy (yellow-green). The flux from the phosphor screen may then be split by means of a dichroic mirror, which diverts most of the actinic flux to the photographic film without significantly reducing the luminance sensed by the observer.

6.2.4 Miscellaneous Filter

Prisms of birefringent material may be used to produce polarized light. A stack of dielectric plates inclined to the wavefront at Brewster’s angle can act as a polarizer.

More often, polarizer construction is based on another phenomenon. A series of long thin conductors, placed parallel to each other, acts as a polarizer, provided the length of the conductors is of the order of magnitude of the wavelength, at least, and their width (and spacing) is considerably less than that. When electromagnetic waves are incident on such a device, the component whose electric field oscillates parallel to the conductor length will set up currents in the conductor and be reflected or absorbed, whereas the transverse component will be affected only slightly.

Such a series of conductors, in the form of a wire grid, was used already to determine the polarization of radio waves. Later such grids were used on microwaves, and even on radiation in the far IR.

By using an ingenious technique (evaporating a thin metallic layer onto a blazed diffraction grating at grazing incidence), such conductor arrays have been obtained sufficiently fine to make them useful even in the near IR region.

In general, polarized light passing through birefringent material changes its state of polarization. The rate of change is proportional to the wave number of the light, inversely proportional to the wavelength, when dispersion effects are neglected. A number of filters based on this fact have been developed.

Birefringent filters have been used in conjunction with a diffraction grating, for tuning a dye laser and spectral widths of about 1pm have been obtained with electrooptic tuning. When a birefringent filter is used inside a laser cavity, the resulting multiple passage of the wave effectively sharpens the transmittance spectrum. In such applications it may suffice to insert a birefringent plate at Brewster's angle, so that the crystal surfaces, act as polarizers and the external polarizing plates may be dispensed with. Tuning may then be accomplished by simply rotating the plate in its plane.

The phenomenon of residual rays can be used for filtering, especially in the IR. In crystals with almost ionic bonding, incident radiation tends to set up vibrations between pairs of neighboring positive-negative valence atoms. These vibrations exhibit resonance effects and at the resonant frequency a large portion of the incident wave field is absorbed in a thin surface layer and converted into mechanical vibrations. These vibrations, in turn, generate radiation (at the same frequency), which appears in the form of enhanced reflectivity. The heightened reflectivity does extend over a considerable wavelength band; but this band can be effectively narrowed by using multiple reflections.

Thin metal films, opaque in the visible part of the spectrum, may have regions of high transparency in the far UV and may, therefore, be useful as filters there.

Since these filters are generally used below 0.1μm, the substrate absorption becomes a major problem. Cellulose nitrate and related materials have been used as substrates, but with proper precautions many metal film can be used totally unbacked.

6.3 Integrated Optics

In many applications light waves can replace electric waves ranging from radio to microwave frequency. The potential advantage of this replacement lies in the great reduction in component dimensions that, in principle, is possible. Light waves are shorter than microwaves by a factor of 10^4, so that miniaturization by 10^8 in cross-sectional area, 12 full decades in volume, is conceivably possible.

The advantages of miniaturization are manifold: (1) Reduced materials costs (e.g., in monocrystalline materials). (2) Reduced manufacturing costs. (3) Reduced packaging costs. In some cases, this may make feasible certain devices that otherwise would be totally impractical. (4) Reduced influence of environment, such as vibration. (5) Speeding up of energy and information transmission due to the shorter distances. This may be important in its own right (as in computer applications) or in terms of the increased bandwidth made feasible thereby. (6) Reduced energy requirements. (7) Facilitating high-energy densities necessary in some applications.

For these potential advantages to be realized, ways must be found to produce highly compact components: (1) Optical waveguides whose cross-sectional dimensions are indeed of the order of magnitude of an optical wavelength. (2) Light sources and detectors of similar dimensions. (3) Light modulation and steering devices of a size consistent with (1) and (2).

Ideally, all the components of an integrated optics subsystem, including light sources, light guides, modulators, couplers, and detectors, would be developed integrally on a single substrate, resulting in monolithic integrated optics. At present, gallium arsenide is the most promising "universal" substrate. On the other hand, it may be desirable to combine components of various substances in a hybrid integrated optics device. For example, light modulators would be made of a material chosen for its large electrooptic coefficient, and the guide, for its high transmittance. It may then be possible first to deposit the localized components on the substrate and then to form an overall guide layer over them. The light may be transferred from guide to component and back again by an appropriate choice of relative refractive index or by the proper shaping of the microscopic interfaces.

Many of the techniques developed in semiconductor electronics industry, such as photolithographic masking, crystal growing, ion implantation, etc., are suitable in the manufacturing of integrated optics components and systems, as well.

6.3.1 Manufacturing

The channels of integrated optics may be produced either by the deposition of one material onto a substrate of some other material, or by modifying volume elements on the surface of the substrate, or inside it, by physical or chemical processes.

Sputtering refers to removal of material from a surface by means of ion bombardment. Frequently this term includes the deposition of the sputtered material onto another surface. Sputtering is a popular method for generating the thin films used in integrated optics components because the resulting films tend to exhibit high transparency.

Sputtering is a rather slow process. Films grow at a rate of approximately 2~3nm/sec; but their uniformity can readily be maintained to within 1% over a 20cm area.

Layers of metal oxides may be deposited by sputtering from metal targets if the atoms are permitted to react with oxygen atoms in the discharge. This technique is called reactive sputtering.

Complex polycrystalline layers can be deposited by chemical vapor deposition, by permitting two chemicals to react in the vapor phase, with the product forming a deposit on the substrate.

Varnishes and epoxy resins may be suitable media for waveguides in integrated optics. The substrates may be dipped into the material and withdrawn slowly. They may be withdrawn rapidly, but then the excess material must be permitted to run off. Spinning, too, may be used to spread the material more uniformly.

This technique is rapid and the film composition, including doping, may be readily controlled.

However, it is difficult to obtain good uniformity.

Lightguiding films may also be grown epitaxially on compatible crystal substrates, with a somewhat lower refractive index. The required difference in refractive index may be due to one of the following: (1) A difference in composition and basic structure of the two crystals (heterostructure). (2) Free charge carriers, whose presence controls the refractive index.

A typical example of heterostructure integrated optics is a gallium-aluminum arsenide layer grown onto a gallium arsenide substrate by liquid-phase epitaxy.

The structure based on differences in free carrier concentration is illustrated by a layer of relatively pure, intrinsic gallium arsenide, grown onto a n-doped GaAs substrate from the vapor phase. The higher density of free charge carriers in the substrate lowers its refractive index, so that light in the intrinsic layer may be trapped and guided.

The junction between two types of GaAs may also serve as a semiconductor junction laser in integrated optics.

Turning now to methods for modifying existing layers in a substrate, we mention first the diffusion of donor-producing (n-type) atoms into a substrate doped with acceptor-producing (p-type) materials, or vice versa.

Alternatively, both semiconductor crystals and glasses can have their surface layers modified by bombarding them with protons or light ions. In crystals, such bombardment creates traps and hence reduces the density of free carriers; this increases resistivity and refractive index. The trap creation is thought to be due to displacement of lattice atoms disrupting the crystal structure.

Frequently it is desired to deposit a certain layer over a limited area or to remove material from a certain precisely defined region. This may be accomplished by masking, executed photographically or electrographically; this technique was developed for use in semiconductor electronics. The substrate is coated with a layer of photoresistive material. It is then exposed by an UV or optical image, or by an electron beam. This exposure changes the physical characteristics so that the subsequent processing removes only the exposed regions (positive) or the unexposed regions (negative material). After the deposition of the desired layer (or removal of the unwanted material), the remaining photoresist is removed.

In some processes, a metallic mask is required. In these, the substrate is first coated with a thin metallic film, for example, by evaporation in a vacuum, and the photoresist is deposited over the metal film. Subsequent exposure and processing leaves the metal film exposed, except in the regions to be covered by the final mask. Now the uncovered metal film can be removed, possibly by an acid treatment. The remaining photoresist is then removed, leaving the mask in the form of a thin metallic film.

When shaped individual channels are desired, the strip intended for the channel may be masked by a quartz fiber bent to the desired shape. The desired thickness of material is now removed by sputtering from the surface, remaining only in the region covered by the fiber, and the desired channel results.

6.3.2 Propagation in a Waveguide

Guiding light within thin film and along small channels is a major function in integrated optics.

These films and channels are then said to serve as optical waveguides, and we present here a brief survey of the basic theory underlying this process. Qualitatively, the phenomenon may be understand in terms of light rays that are totally reflected at the interface between the guide and the substrate or other matrix in which it is embedded. The condition for total reflection is simply that the angle of incidence of the ray on the interface equal, or exceed, the critical angle θ_C, defined by

$$\sin\theta_C = n'/n \tag{6.1}$$

where n, n' are, respectively, the refractive indices of the media on the incidence side and outside of the interface.

When the thickness of the guiding layer far exceeds the wavelength of the light, and there is a substantial difference between the refractive indices, the performance can be analyzed accurately in terms of rays. When these conditions are not met, only certain rays (or field modes) are possible. These are found by means of Maxwell's equations with the appropriate boundary conditions. Most of the phenomena may, however, be explained in terms of rays, multiple beam interference and Fresnel's equations.

6.3.3 Coupling of Wave Energy

The transfer of light from one component to another is done via radiation coupling. This occurs when light from a radiation source is introduced into an optical waveguide and when it is transferred from the waveguide to the detector. It may also occur between light guides, transferring light from one guide to another. A number of mechanisms are available for accomplishing this.

The most obvious way of inserting light into a waveguide is by simple illumination of guide edge, called "end fire" coupling. However, for this insertion to be efficient, the distribution of the illumination must match the amplitude-squared distribution in the guide. Simultaneously, the illumination must be limited in angular distribution, again to match the modes to be excited in the guide. This limitation precludes substantial demagnification of any source not highly collimated. Consequently, efficient coupling is feasible only from lasers. Since the zero-order modes of slab and rectangular waveguides resemble the Gaussian distribution often obtained from laser source, the coupling could, in principle, be made quite efficient. In practice, however, the method is very difficult, because the microscopic dimensions of the guide demand enormous precision on part of the associated optics.

More efficient methods of coupling are based on techniques in which electromagnetic flux runs parallel to the guide and gradually leaks into it.

Tapered film can be used to feed laser light into a waveguide, as well as to transfer light from one guide to another in hybrid integrated optics circuits. The prism coupler consists of prism, and close proximity to, the waveguide layer, a thin film in our example. The light enters the prism and is partially reflected at the base, which runs parallel to the film surface, or almost so. Part of the incident flux, however, penetrates the airgap and enters the film to excite in it guide modes. The grating coupler works on a principle similar to that of the prism. The light is incident at an angle such that the diffraction angle corresponds to that of one of the permitted zig-zag directions.

Alternatively, the condition may be described in terms of the periodic field variations of the wavefront obliquely incident on the grating, as modified by the grating complex transmittance: for efficient flux transfer one of the harmonic components of this product must match the phase of a permitted mode of film.

The type of coupling described in our discussion of the prism coupler, occurs also whenever two optical waveguides are arranged so that their mode fields overlap. This is called directional coupling. For example, if two identical waveguide channels run parallel and in close proximity to each other, the wave propagating in one of them will slowly be transferred to the neighboring one. Upon completing the transfer to the second one, the wave will begin to be transferred back to the first: the energy will oscillate back and forth between the two guides.

6.3.4 Components

Having surveyed the major processes utilized in integrated optical systems, we now turn to a discussion of the chief components used.

- **Waveguides**

An optical waveguide consists of an extended region of relatively high refractive index and low attenuation surrounded by media of lower refractive index, so that electromagnetic waves can propagate in it with reasonable small absorption and radiation losses. The difference in refractive index may be obtained by making the guide of a change in free carrier concentration.

Waveguides may be classified according to the shape of their cross section. The simplest such guides are in the form of films and are effectively limited only in one dimension. These are called slab or planar guides. When limited in width as well, they are called channel waveguides. Those without substrate are called fibers.

Channel waveguides may be deposited above a substrate; these are called ridged or strip guides. They may also be embedded in the substrate. A third alternative, the strip-loaded planar guide, is a hybrid of the ridged channel and the planar waveguide.

- **Light sources**

In principle, any light source can be used in integrated optics. In general, however, effective coupling is difficult. For efficient coupling, a small source area must be combined with directional concentration, and often monochromaticity, too, is important. Lasers combine all these requirements and are, therefore, the most popular light sources used in integrated optics. Also, strong coherence is necessary if a single mode is launched, and this, too, is relatively readily available from a laser. Light emitting diodes (LED) satisfy the small area and monochromaticity requirements and, therefore, are an occasional alternative to the laser.

The choice among the various types of laser available is dictated by the requirement of small size. This leads to the solid-state and semiconductor lasers as prime candidates. For ease in pumping, the injection-pumped semiconductor laser appears to be the most important light source in integrated optics. Such a laser can be coupled to the waveguide by any means available for radiation coupling.

- **Lenses and prisms**

Lenses and prisms in "two dimensional" versions can be incorporated into integrated optics systems. In one technique, the lens or prism is deposited on the substrate in the form of a thin cross-sectional slice of the prototype, generally via a mask. A coupling film is then deposited over them. This permits coupling the wave into the prism or lens and from one into the other, via the coupling film. The edges may be either sharp or tapered. An isosceles right triangle, with the hypotenuse normal to the axis, has been suggested as a "two-dimensional corner-cube", possibly to serve as a reflector in an integrated optics laser cavity.

- **Modulators**

The light passing through the integrated optical system may be modulated in accordance with an applied electrical signal by means of any of the basic techniques—electrooptic, magnetooptic, or acoustooptic—used for high-frequency modulation in the usual optical system.

If an electric field is applied to an optical waveguide made of electrooptically susceptible material, birefringence is introduced, in general. Consider an electric field applied normal to the wavevector of the light. If the light is polarized in the direction of the applied field, or normal to it, the induced birefringence will appear simply as a change in refractive index. If the light is linearly polarized in any other direction, or elliptically polarized, its ellipticity will change as it passes through the guide. Either of these two effects may be used to produce amplitude modulation.

The periodic index modulation can also be obtained by means of acoustic waves. As surface waves, these modulate the thickness and as compressional waves the refractive index. Both of these form gratings coupling a frequency change into the lightwave propagating along the guide and, hence, may switch it into another mode. This new mode may be lost to (or uncoupled from) a parallel guide. In planar slab guides, the grating may change the direction of propagation of the guided beam and hence switch it from one detector to another.

A magnetooptic effect, Faraday rotation, too, has been used for modulating the flux in an integrated optics system. The modulation was effected by switching flux from the TE to the TM mode and vice versa.

- **Detectors**

The detector used in conjunction with an integrated-optics device may, of course, be external with the light coupled into it. Ideally, however, the detector should be integral with the system as a whole. The coupling can then be end-on, preferably by having the detector butt the guide; or a coupling layer, with its tapered edge, may be used in hybrid integrated optics systems.

A number of detectors that have been built integrally in this manner are listed below: (1) Silicon p-i-n diodes have been diffused by the usual techniques onto a silicon substrate. The associated lightguide was a film of glass sputtered onto the substrate with a layer of SiO_2 serving as an isolation layer. Coupling to the diode was via a tapered guide. Such a diode, with its intrinsic, light-sensitive region buried some micrometers below the surface, is ideal for use with embedded waveguides. (2) Schottky barrier diodes are more suitable for use with surface guides, in conjunction with a tapered coupling film. (3) Detectors may also be grown epitaxially into a hole etched into the waveguide region.

If the detector could be made of the same material as the guide, this would offer great advantages in manufacturing economy. The obvious difficulty here is that, as a waveguide, the material must be transparent and, as a detector, it must absorb light efficiently.

Several techniques may be used to accomplish this: (1) If the material is used near the bandgap, an electric field may induce electroabsorption and this may create a photodiode in a heterostruture waveguide. (2) Ion implantation may change the guide material locally into a photodetector. (3) Diffused impurities may be used to convert a portion of the guide into a detector.

6.4 Photoelectric and Thermal Detectors

To detect light we may, in principle, use any action it exerts on a substance. In a solid the electrons are distributed in energy bands separated by forbidden gaps. The phenomena of interest to us occur between the valence band where electrons are almost immobile, the conduction band where they are quite mobile, and the semi-infinite "band", which we may call "freedom", where electrons are no longer bound to the original solid. This band exists only beyond the surface and overlaps the conduction band, at least in part.

If the electron receives an amount of energy sufficient to raise it to the vacuum level (the bottom of the "freedom" band), it may escape the solid. Once freed, the electron may be transported to another structure for detection. The freeing of electrons from a solid by incident radiant flux is called the photoemissive effect.

When the absorbed radiation transfers the electron only into the conduction band, this is called the photoconductive effect. The direct use of this effect is important, but frequently it is used in conjunction with semiconductor junctions that permit the manipulation of the motion of the freed carriers.

The excitation of an electron is a rapid process requiring only about 10^{-12} sec. In photoemission, this excitation, together with the subsequent rapid electron transport, essentially completes the detection process. Consequently, the response times of photoemissive detectors are generally very short (of the order of nanosecond) making them suitable for high-frequency operation and the observation of short-duration events. In photoconductive mechanisms, the operation may depend critically on carrier life times, which are quite long, often measured in milliseconds.

Not all incident photons participate in the above processes. Some are reflected or transmitted, others are absorbed in phonon processes and still others have the electrons, freed by them, absorbed. The fraction of photons that participate in any detection process is called the quantum efficiency of the process.

6.4.1 Phototubes

To obtain useful photoemitting cathodes, a relatively low work function must be provided. Clean metal surfaces are relatively poor photoemitters. On the other hand, when monomolecular layers of alkali metal ions are absorbed on such bases as silver or tungsten, a far better response is obtained. This is, presumably, due to the formation of an electrical double layer on the metal surface. Further improvements may be obtained by partial oxidation of the alkali layer.

However, to obtain high efficiencies, a low work function does not suffice; the cathode must be such that useful photoexitation can take place also within the material up to the depth of several tens of nanometers. Once they have been excited into the conduction band, electrons must be able to reach the surface with a major portion of their energy intact. This is much more likely in a dielectric or a semiconductor than in a metal. On the other hand, there must be enough electrical conductivity to permit the neutralization of the charge accumulated due to the photoemission; otherwise, the positive charge on the cathode will inhibit the escape of photoelectrons, resulting in fatigue effects.

Efforts to satisfy all these requirements have resulted in a number of highly efficient photocathodes with various efficiencies and spectral responses. The two notable phocathodes were the silver-oxygen-cesium (Ag-O-Cs) and the antimony-cesium (Sb-Cs) cathodes. Comparing these, we note that the Ag-O-Cs cathode has the lowest work function of any commercially available cathode and can, consequently, be used well into the infrared, to beyond 1.1μm. Its peak sensitivity, on the other hand, is only one-tenth of that attainable in the more popular Sb-Cs cathodes. Due to its low work function, its dark emission is several thousand times as high as that the Sb-Cs cathode. These two factors limit its usefulness at very low illumination levels. The Sb-Cs cathodes are usable from the ultraviolet to approximately 0.7μm and, in the visible region, have highest sensitivity commercially available.

Photocathodes may be made either opaque or partially transparent. The later form is usually used with light incident on one side of the cathode and electrons emitted on the other. This permits a tube construction with better optical coupling characteristics. On the other hand, the cathode thickness becomes critical, if the cathode is too thin, it will not absorb enough photons; if too thick, the photoelectrons will not be able to escape for collection at the anode. Also, the required electrical conduction is difficult to obtain with the very thin transparent cathodes, so that conducting tin-oxide layers are occasionally provided under the cathode layer.

The cathode spectral response is somewhat dependent on the angle of incidence. Especially in the thinner cathodes increased obliquity increases the absorption of red radiation resulting in a slight shift of the response toward the long wavelengths.

The simplest device employing photoemission is the vacuum photodiode. This consists of a photoemissive cathode and an electron-collecting anode, enclosed in an evacuated envelop that permits the entry of the radiation to be measured.

Generally, the photocathode is made large to facilitate the collection of a maximum of light flux. In photodiodes having opaque cathodes, the emission of electrons is toward the incident light. The collecting anode must then be on the same side. It is, therefore, generally made in the form of a relatively thin wire to minimize the blockage of the incident light.

6.4.2 Gaseous Amplification and Gas Photodiodes

To increase the current output of a photodiode without adding a significant amount of noise, internal current amplification must be employed.

Consider an electron, accelerated by an electrostatic field, traveling through a gas. There is a certain probability that is will collide with a gas molecule. If, at the time of the collision, the electron has enough energy to ionize the molecule, this collision may produce a second free electron (plus a positive ion). We now have two electrons being accelerated by the electrostatic field; each of these may participate in another ionizing collision to double, again, the number of free electrons in the current. This phenomenon is known as gaseous current amplification and is used to increase the output of photodiodes.

Since gas ions contribute significantly to the total current, both as positive ions and via secondary electron generation, and since these ions move very slowly compared to electron velocities, the response of gas diodes is considerably slower than that of vacuum diodes.

Often the output of a phototube must be amplified for measurement and detection. Invariably, such amplification adds noise to the signal. The amount of noise added, however, depends on the method of amplification. A relatively low-noise amplification process is that of electron-multiplication, and therefore this is frequently used in phototubes. Tubes employing such electron multipliers are called multiplier phototubes.

When an energetic electron strikes a solid, it excites electrons internal to the solid and some of these excited electrons may escape. The number of electrons thus freed depends on the energy of the incident electron and on the material bombarded; the energy of the true secondary electrons is approximately 2eV and does not change significantly either with the bombarded material or with the energy of the incident electron. The secondary emission ratio is defined as the ratio of the total number of emitted electrons (including scattered incident ones) to the number of electrons incident.

Multiplier phototubes consist of an evacuated envelope containing an electron-emitting cathode, a number of electron-multiplying electrodes called dynodes (or a channel electron multiplier) and an electron collecting anode. The construction must allow the light to strike the photocathode and permit the collection of most of the photoelectrons onto the first dynode at an energy appropriate for the desired multiplication ratio. Successive dynodes and the anode must be placed and electrified such that they similarly collect the electrons emitted by the preceding dynode.

6.4.3 Photoconductive Detectors

Photoconductive detectors fall into two major categories: (1) Homogeneous detectors where a carrier, freed anywhere inside the material is drawn out by an externally applied field. (2) Junction detectors, where the carriers are generated in the neighborhood of a p-n junction to be separated by an internally generated field.

In the junction detectors, with the exception of the avalanche diodes, there is no such gain. On the other hand, very short time constants are attainable.

It is convenient to divide the homogeneous detectors into intrinsic and extrinsic types. The extrinsic detectors are more difficult to make, because of the critical nature of the doping levels. Also, their responsivity is lower because of their lower absorption. On the other hand, they provide far greater flexibility.

The most commonly used intrinsic photoconductors are cadmium sulfide (CdS) in the visible region and lead salts and indium antimonide (InSb) in the near infrared.

To obtain sensitivity extending further into the IR, carriers with lower excitation energies are required. These may be provided by means of "doping" i.e., the substitution of atoms of different valence for some of the matrix atoms. In principle, both silicon and germanium are suitable, and both positive and negative doping may be used. In practice, however, positively doped germanium is used almost exclusively.

When materials with very low excitation energy are used, there is a strong tendency for thermal excitation to prevail and to mask the photo-excitation of interest. This can be prevented only by cooling; the cooling requirements will be the stricter the further into the IR the sensitivity extends.

Photodiodes may be used to overcome the limitations of phtoconductors as inherent in their time constants. In addition to this advantage, photodiodes provide the option of operating in a photovoltaic mode: they may be used as active power sources, where the power generated is proportional to the incident radiation. When used in this configuration, they may be used to convert electromagnetic flux, such as sunlight, directly into electrical power. Diodes constructed for this purpose are called solar cells. The highest detectivities are attained with reverse bias, and diodes in detection applications are generally operated with such a bias.

To overcome the lack of internal gain in the ordinary photodiode, it may be operated at a voltage sufficiently high to provide avalanche breakdown or a dual junction (transistor) form may be used.

Diode junctions may be of the p-n type. Alternatively, they may consist of a metal-semiconductor interface called the Schottky barrier or the classical point-contact junction. All of these may exhibit both photovoltaic and rectifying effects.

6.4.4 Image Detectors

Up to this point, we have discussed photoelectric detectors that measure total flux and do not respond usefully to change in flux distribution. In conjunction with scanning devices, mechanical or optical, such detectors are capable of investigating the distribution of flux in an image. There are, however, other photoelectric detection devices that are capable of transforming an input irradiation distribution into a new one or into a sequence of electrical signals. The former type of device is usually called an "image tube" and the latter an "image pick-up tube", (television) camera tube or "signal generating tube". A more appropriate nomenclature might be optical-optical and optical-electrical image transducers.

In many of these devices, the optical image is first converted into an electron image, that is, an electron cloud whose density distribution follows that of the optical irradiation distribution.

In image tubes including cathode ray display tubes, image intensifier tubes, and television "camera" tubes, electron optics are used for three essentially different functions: (1) Imaging. An electron density distribution in one plane is reproduced in another plane. Here the exact distribution of electrons is significant. (2) Beam concentration. The electrons in beam are concentrated into a small cross section, with no regard to the exact distribution within this cross section. (3) Deflection. The electron beam is deflected and guide to various desired locations.

Imaging is the essential function in image intensifier tubes and in the "write" stage of television camera tubes. Beam concentration, in conjunction with deflection, is used in cathode ray tubes and in the "read" stage of television camera tubes. All of these functions can be performed either electrostatically or magnetically.

6.4.5 Thermal Detectors

The thermal detectors respond due to a change in temperature which is a function of the absorbed energy and independent of its wavelength; their spectral response is determined solely by their spectral absorptivity and is therefore generally much more uniform than that obtained with a simple photon detector.

Thermal detectors are mostly elemental, that is, they measure the total radiation incident on their sensitive region, regardless of its distribution. Thus, scanning must be used to observe an irradiation distribution. However, image-forming thermal detectors, too, have been developed.

In principle, any substance having a measurable physical characteristic that changes in a predictable manner with temperature can be used as a thermal radiation detector. In practice, there are, however, primarily only four types of such detectors in use: (1) the bolometer, (2) the thermocouple, (3) the pyroelectric, and (4) the pneumatic detector.

6.4.6 Comparison of Detectors

When they cover the required spectral range, photon detectors will be found, as a rule, to be more sensitive than thermal detectors and will therefore be preferred, except when they must be rejected for specific reasons. Such reasons may be the price, the cooling, which is required for all the photon detectors in the far infrared or, possibly, a limitation in the dynamic range. When a very broad, or uniform, spectral response is required, the thermal detectors may be superior.

Photoemissive (vacuum tube) detectors are limited almost exclusively to the near IR range of the spectrum and below. Because of the feasibility of electron multiplication, their gain and overall detectivity is there generally superior to those obtainable with photoconductive detectors. However, when operation is required only at higher levels of illumination, where these advantages are not effective, the more convenient photoconductive detectors may be preferable even in this spectral range. Especially in applications where it is desirable to operate without an external power supply, the photovoltaic or photoelectromagnetic modes of operation of the photoconductor will make it the preferred device.

It is found that the photodiode is superior to the homogeneous photoconductor at low light level (milliwatts). On the other hand, because of the high intrinsic capacitance of the diode, its output circuit must be matched to the operating frequency and, with broad-band operation, the efficiency will be lower in diodes than in the homogeneous photoconductors.

In terms of responsivity, homogeneous photoconductors are superior because of the high gains of which they are capable, whereas the highest gain-bandwidth products are obtained in avalanche diodes.

Words and Expressions

accumulate	积累，堆积
acid	酸
actinic	光化学的
adjacent	邻近的，毗连的
airgap	空隙
alkali	碱性的
anisotropy	各向异性
anode	正极，阳极
antimony	锑
apochromatization	复消色差，消多色差
attenuation	衰减
avalanche breakdown	雪崩击穿
bandgap	能带隙
batch	一炉
birefringence	双折射
blazed diffraction grating	闪耀衍射光栅
bolometer	辐射热测量仪
bombardment	轰击
butt	拼接
cadmium sulfide	硫化镉
carrier concentration	载流子浓度
casting	铸造
cathode	负电极，阴极
cellulose nitrate	硝酸纤维素
cesium	铯
chromaticity	色度
colloidal	胶状的，胶质的
colorimetry	色度学
compact	紧凑的，紧密的，简洁的
compatible	相容的，兼容的
conceivably	令人信服地
conduction band	导带

conjunction	结合
contour	外形 ；描……的轮廓
corner-cube	角隅棱镜，三面直角棱镜
cross-section	横截面
crucible ['kru:sibl]	坩埚
crystalline	水晶的
demagnification	缩小
deposit	沉淀
dichroic	二向色的，二色性的
dielectric	电介质
dip	浸，蘸，沾
discharge	放电
dispense with	免除，无须
disrupting	中断，分裂，瓦解，破坏
dissolve	（使）溶解，溶化
donor	施主，供体
doping	掺杂
dye	染料，给……染色
dynamic	动态的，动力的，动力学的
dynodes	倍增电极
efficacy	效能，功效
electrify	使充电，使通电，使电气化
electrodes	电极
elemental	基本的
embed	把……嵌入，扎牢，使深留脑中
emulsion	感光剂
epitaxially	外延地
epitaxy	外延，取向生长
epoxy resin	环氧树脂
equilibrium, equilibria（复）	均衡
etch	蚀刻
evacuated envelop	真空包壳
extrinsic detector	非本征激发的探测器
facilitate	使容易，使便利
fatigue effect	疲劳效应
feasible	可行的，可能的，可用的
flexibility	灵活性，弹性
fluorite	萤石，氟石
flux	通量
fused silica	熔融石英

fusion	融化，熔融
gallium arsenide	砷化镓
gallium-aluminum arsenide	镓铝砷
gaseous	气体的，气态的
gelatine	白明胶
germane	相关的
grazing incidence	掠入射
grind	碾，磨
halide	卤化物的
hetero-	[词头]杂，异，不同，不均一
heterostructure	异质结构，异晶结构
homogeneous	同质的，均匀的
hybrid	混合的，混杂的
hypotenuse	直角三角形之斜边
immobile	固定的
impact	冲击，碰撞
implantation	注入
impurity	杂质，混杂物
in accordance with	与……一致，依照
in conjunction with	与……协力
in turn	依次，轮流
incorporate	结合，合并，加入
indice	指标
indium antimonide	锑化铟
infrared	红外线的
ingenious	精巧
inhibit	阻止，禁止
inorganic	无机的
intact	未扰动的
integrated optics	集成光学
interatomic	（同一分子中的）原子间的
interpose	插入，放入
intrinsic	固有的，内在的，本质的
intrinsic detector	本征探测器
ionic bonding	离子键
isotropic	各向同性的
joint	接合处，接合点
junction	接合点，接合，交叉点
lattice	晶格，点阵，网格
lead salt	铅盐

leak	漏，渗
localized	局部的
luminance	亮度
manifold	多方面的，多种形式的
manufacture	制造
mechanical	机械的，力学的
melt	熔体，熔化物
microscopic	极小的，显微镜的
miniaturization	小型化
miscellaneous	各色各样混在一起的，混杂的
mobile	活动的
mobility	活动性，灵活性，迁移率，机动性
modify	修正
monochromaticity	单色性
monolithic	单片的，单块
motivate	激发
mount	安装，装备
multiplier phototubes	光电倍增管
neutralization	中和作用
nomenclature	术语
obliquely	斜
obliquity	倾斜度
occasional	偶然的，特殊场合的，临时的
optically active	旋光的
optical plastics	光学塑料
orientation	方位，方向
oscillate	摆动
penetrate	穿透，渗透
phonon	声子
phosphor	荧光体
photoconductive detector	光电导探测器
photoconductive effect	光电导效应
photodiode	光电〔敏、控〕二极管
photoelectric	光电的
photoemissive effect	光电发射效应
photolithographic	光刻法的
photoresist	光刻胶，光阻材料
photoresistive	光电导的，光阻的
phototube	光电管
photovoltaic	光生伏特的

pneumatic	气动的
polaroid	偏振片
polishing	抛光，磨光
polycrystalline	多晶的
potential	潜在的
precautions	预防
precision	精确（性），精密（度）
preclude	排除
prevail	流行，盛行，获胜，成功
prism	棱镜；（结晶）柱
projection	投影，投射
promising	有希望的，有前途的
pronounced	显著的
proton	质子
prototype	原型
pyroelectric	热电的
quartz	石英
quench	熄灭，淬火
reactive sputtering	反应溅射
rectifying effect	整流效应
residual	剩余的
resonant	洪亮的
reverse bias	反向偏置
ridge	（山）脊，峰
rigid	坚硬（固）的，固定（连接）的，严格的
schottky barrier diodes	肖特基势垒二极管
scratch	擦伤，刮伤
silica glass	石英玻璃，硅玻璃
silicate	硅酸盐
silver halide	卤化银
slide	幻灯片
sophisticated	复杂的，尖端的
specific gravity	比重
spin	自旋，旋转
sputtering	溅射
stack	整齐的一叠
steering	转向，操纵
subcategorie	子范畴，分类
substance	物质
susceptible	易受影响的

suspension	悬浮液
synthesize	综合，合成
tapered	锥形的
terminology	术语学
tetrahedral	四面体的
tetrahedraon, tetrahedral（复）	四面体
thermocouple	热电偶
thin-film	薄膜
tin	锡
transducer	传感器，变换器
transition	转变，转换，跃迁，过渡
transparency	透明片，幻灯片
tungsten	钨
tuning	调节，调音
ultimately	最后，基本上
valence	原子价
valence band	价带
varnish	清漆
via	通过
vibration	振动
vicinity	邻近，附近，接近
waveguide	波导
wire grid	线栅
withdraw	取出，提取，分离
work function	功函数

Grammar 专业英语翻译方法（六）：状语从句的译法

状语从句根据它的含义可分为 9 类：时间、地点、原因、条件、方式、目的、结果、让步、比较或对比。在汉译时，有些状语从句可以改性，译为其他状语从句，如时间或地点状语从句改译为条件状语从句等。

—As soon as the current has reached a constant value, the induced electromotive force disappears.

电流一达到恒定值，感生电动势就消失。（时间）

—When steam passes over red-hot iron, iron oxide and hydrogen form.

当蒸汽从炽热的铁上通过时，就会生成氧化铁和氢气。（时间转化为条件）

—The materials are excellent for use where the value of the workpieces is not so high.

如果加工件价值不高，最好使用这些材料。（地点改为条件）

—The mechanical efficiency of an engine is always less than 100 percent because losses in form of friction are inevitable.

一台发动机的机械效率总是小于 100%，因为摩擦损失是不可避免的。（原因）

—Now that you are all ready, we had better start the engine at once.

既然你们都准备好了，我们最好立刻就启动发动机。（原因）

—Should something go wrong, the control rod would drop.

万一发生什么事故，控制杆就会落下来。（条件）

—Workers should operate the machines as the instructions specify.

工人们应该按照说明书的规定操作机器。（方式）

—Steel parts are usually greased lest they should rust.

钢制零件通常涂以油脂，防止生锈。（目的）

—The induced emf is in such a direction that it opposes the change of current.

感应电动势的方向是阻止电流发生变化的那个方向。（结果改译为表语）

—Although the factory is small, its products are of very good quality.

这座工厂虽小，但是它的产品质量却很好。（让步）

—The thicker is the wire, the smaller is the resistence.

导线越粗，电阻越小。（比较）

Part 7 Applied Techniques

In this part, we will introduce some applied techniques including optical thin film technology, photolithography, biophotonics, 3D Display Technology, Infrared Detection Technology.

7.1 Optical Thin Film Technology

Optical thin films are widely used today in many diverse applications to control the way light is reflected, transmitted, or absorbed as a function of wavelength. They can be grouped into two major categories based on the application. In the first, the light travels parallel to the plane of the substrate with the films acting as wave guides in the emerging field of integrated optics. Here light signals could replace electrical signals in applications such as communications and computers. In the second application, the light travels perpendicular to the film plane for use as antireflection coatings, edge filters, high efficiency mirrors, beam splitters, etc.

The term "thin" is used to indicate a layer whose thickness (perpendicular to the substrate) is the same order of magnitude as the wavelength of interest, and the extent (parallel to the substrate) is a very large number of wavelengths. Typical layers might range in thickness from 80nm in the visible to twenty times that in the infrared. Filters are composed of a stack of such layers, alternating between high and low refractive indices, with typically 20~40 layers, although in some cases they may have one hundred layers or more. Thin film filters operate by interference of the light reflected from the various layers as the light passes through perpendicular to the substrate.

7.1.1 Design of Optical Thin Film

When light strikes a film, it can either be reflected, transmitted or lost to absorption or scattering. If we consider a light beam incident on a homogeneous parallel-sided film, the amplitude and polarization state of the light transmitted and reflected can be calculated in terms of the angle of incidence and the optical constants of the three materials involved, as shown in Fig.7.1. Light from the incident medium of refractive index N_0 passes through film material of index N, and enters substrate material of index N_s. The incident medium is often air with an index of refraction assumed to be equal to 1.0, the index of vacuum. The film and substrate materials can be transparent or absorbing in which the optical constant (or complex index of refraction), N, is given by

$$N = n - ik \tag{7.1}$$

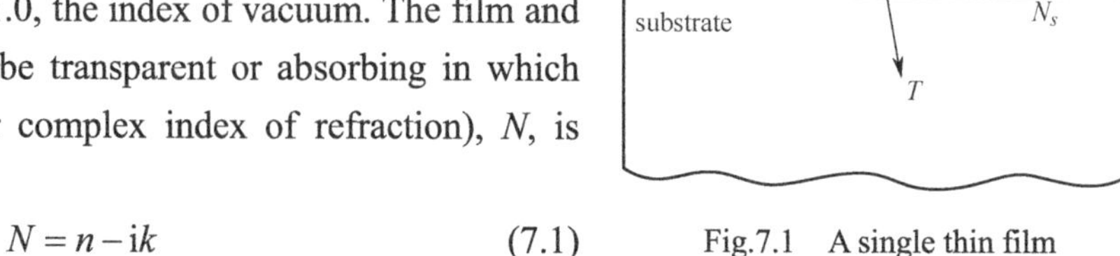

Fig.7.1 A single thin film

where n is the refractive index and k is the extinction coefficient. At optical frequencies, $\varepsilon = n^2$, where ε is the dielectric constant of the material. The extinction coefficient is related to the absorption coefficient, α, by the expression

$$\alpha = 4\pi k / \lambda \tag{7.2}$$

where α determines the intensity I transmitted through an absorbing medium by the exponential law of absorption

$$I = I_0 e^{-\alpha x} \tag{7.3}$$

The reflectance and transmittance at the boundaries between these regions can be conveniently expressed in terms of the Fresnel coefficients. When the regions are absorbing, these terms are large and cumbersome but simplify into terms involving only n if the regions are transparent. Thus, if we assume that $k = 0$ and that plane waves strike a plane boundary at normal incidence, the reflectance between two regions is given by

$$R = \left[\frac{n_0 - n_s}{n_0 + n_s} \right]^2 \tag{7.4}$$

For example, the reflectance of uncoated window glass (with n_s=1.52) in air (with n_0=1.0) is 4.3%. The transmittance, T, through the surface would be

$$T = 1 - R = 95.7\% \tag{7.5}$$

and through both surfaces would be 95.7% of 95.7% or 91.7%. By comparison, the transmittance through a single lens of n=1.9 would be 81.7%. Complex systems may have many lenses cascaded together and the losses may total 50% or more. Infrared systems use very high index materials such as germanium which has a reflectance of 85% per surface. In addition to power lost, these reflections cause "ghost" images, thus it is necessary to put antireflection coatings on the lenses.

Antireflection coatings operate on the principle of interference of the light reflected from the front and back surfaces of the films. The optical thickness of a thin film is defined as the index n_f times the physical thickness d_f. In the case where $n_f d_f = \lambda/4$, $3\lambda/4$, $5\lambda/4$, etc., the equation for calculating the reflectance of a single layer as in Fig.7.1 at normal incidence for the wavelength of interest λ_0 simplifies to

$$R = \left[\frac{n_0 - \frac{n_f}{n_s}}{n_0 - \frac{n_f}{n_s}} \right]^2 \tag{7.6}$$

This is a very useful formula for estimating how well a particular single layer coating will do. We can calculate the film index needed for zero reflectance at one wavelength as

$$n_f = (n_0 n_s)^{1/2} \tag{7.7}$$

For a lens of n = 1.9, we get n_f = 1.33 which happens to be the index of refraction in the visible of MgF_2, the most widely used material for single layer antireflection coatings and one of the lowest index materials available. Thus, for glasses with an index lower than 1.9 (which is usually the case) we must either accept a little higher loss for the economy of a single layer MgF_2 coating or go to the additional cost of adding more layers of other materials. Adding more layers also gives the advantage of being able to achieve a lower reflectance over a much broader wavelength range.

The interference effect can be used by building up a stack of alternating high and low index materials to produce many interesting results. Figure 7.2 shows the construction of such a quarter wave stack in which the optical thickness $n_j d_j$ of each layer is again equal to one quarter the wavelength of interest. The multilayer is completely specified if we know n_j, k and d_i for each layer, n_0 for the incident medium plus n_s and k_s for the substrate. Given the angle of incidence θ_0, we can calculate the reflectance R and the transmittance T as a function of wavelength.

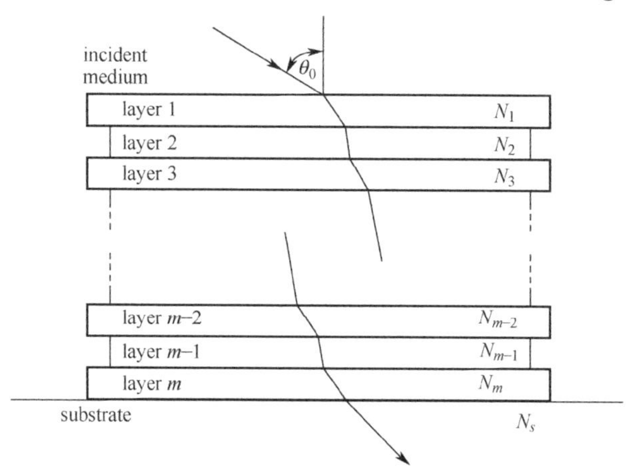

Fig.7.2 A multilayer stack

Figure 7.3 shows the reflectance vs. wavelength of six and twelve layer stacks of Si/SiO_2. They are plotted as a function of λ_0/λ, where the wavelength of interest $\lambda_0 = 4nd$. The multilayer has a characteristic stopband (or high-reflectance region) symmetric about the wavelength λ_0 surrounded by long and short-wave pass regions characterized by many ripples in the passbands. The width of the high reflectance region is determined by the ratio of the high to low index n_H/n_L. The higher the ratio, the wider the stopband. The maximum reflectance depends on the number of layers as well as the ratio n_H/n_L. The ripple in the passband is bounded by an envelope determined by n_H/n_L whereas the number of peaks depends on the number of layers.

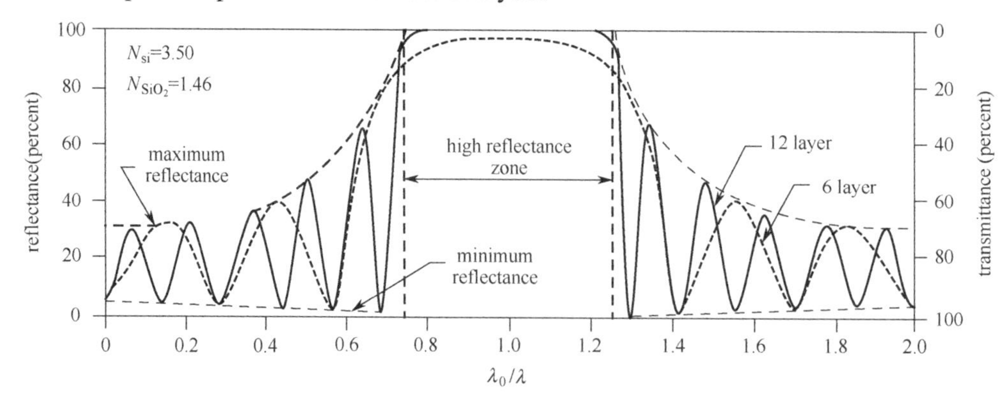

Fig.7.3 Reflectance vs. wavelength for Si/SiO_2 multilayer stacks

By suppressing the ripple on one side of the stopband we can create an edge fitter. A long-wave pass (LWP) filter transmits the long wavelengths and efficiently reflects the short ones. On the other hand, a short-wave pass (SWP) filter transmits the short wavelengths and reflects the long ones. A SWP filter is sometimes referred to as a high-pass filter meaning it passes high frequencies. Likewise, a LWP filter is sometimes called a low-pass filter and care must be used not to confuse these terms. The bandpass filter transmits a relatively narrow range of wavelengths and effectively rejects those on both sides.

Once a design has been made, the wavelength characteristics can be moved in wavelength by changing the layer thicknesses by the same ratio. For example, a LWP fitter with an edge at 400 nm can be shied to 800 nm by making all the layers twice as thick.

As the angle of incidence increases, two effects are seen. First, the effective thickness of the layers becomes smaller which causes the filter to shift to shorter wavelength. Since the effect is inversely proportional to the film index, at large angles the layers are no longer matched at the same wavelength and the spectral shape degrades. The second effect at non-normal incidence is that the spectral characteristic becomes dependent on the polarization of the incident beam. This can be optimized and used to advantage for such components as polarizing beam splitters. But in general, these angle effects are detrimental, especially for a non-collimated beam.

7.1.2 Deposition of Optical Thin Film

There are many methods of thin film deposition. The major technique in use today for the deposition of optical films can be termed physical vapor deposition (PVD) in which the material condenses from the vapor phase onto the desired substrate. There are many different ways the vapor can be produced, but thermal evaporation (or vacuum evaporation) is the most widely used technique today. It is carried out in a vacuum to prevent unwanted chemical reactions and to keep the film from becoming porous by incorporating gas during the condensation process. The substrate to be coated is typically placed in a vacuum chamber, pumped down in the range of 1×10^{-6} torr (10^{-9} atmosphere) and heated to 200~300℃. The coating material is then heated until it evaporates or sublimes, travels across the vacuum chamber, then condenses on the substrate.

The condition and cleanliness of the substrate is critical to the quality of the coating. Since the forces which hold the film onto the substrate are all short range, even a few molecular layers of contaminant will result in a film with reduced adhesion. Since the layers are only a few tenths of a micron thick, they will faithfully reproduce any irregularities in the surface and cannot be expected to smooth over imperfections left from grinding and polishing. As a matter of fact, imperfections and especially stains in the glass surface that might be unobservable are greatly enhanced by coating. Once the substrates are cleaned and loaded into the fixtures they should be put under vacuum and coated as soon as possible.

The fixture is generally placed at the top of the chamber with the substrates held in by gravity. To improve thickness uniformity, the holder is often a dome which is rotated about the vertical axis of the chamber. By using uniformity masking, thickness from one area of the fixture to another can be controlled to ±2% or better. Tighter control may often be achieved with planetary tooling where smaller plates are rotated about their axes which are also moved about the chamber axis in a planetary motion. Some argue that the slight improvement in uniformity possible with the planetary is not worth the additional capital cost, complexity, particle generation and reduced throughput.

For vacuum evaporation, a high vacuum on the order of 10^{-6} torr is necessary so that evaporant molecules will travel across the chamber without encountering other molecules. This can be obtained with a variety of high vacuum pumps, the most common being the hot oil diffusion pump. It has been the workhorse of the industry for many years because it is economical, reliable and has a high throughput. If operated properly, it will not contaminate the chamber and substrates with oil. However, if oil contamination is seen, it is most likely from the mechanical pump used to "rough down" the chamber and back up the high vacuum pump. In recent years, cryopumps and turbomolecular pumps have sometimes been used in place of diffusion pumps because of the perception that they will eliminate oil contamination.

Rough pumping takes on the order of 5~10 minutes with high vacuum pumping taking an hour or more. During this time the substrates are usually heated to improve adhesion and film properties. The heaters are either the rod type placed behind the substrates or, more often, quartz lamps placed below the substrates. In both cases, heating is by radiation in the vacuum.

The two most widely used methods of evaporant heating are the resistance source and electron beam gun. In resistance heating, a large current of perhaps a few hundred amperes at a low voltage is passed through a boat or filament to evaporate or sublime the material. The electron beam gun generates a stream of electrons of 6~10kV up to a few amperes. Using magnetic fields, the beam is deflected typically 270℃ to keep the filament out of the path of the evaporant which greatly extends the filament lie. Electromagnets are used to move the beam around on the source material to control what area gets heated, as well as to raster the beam to avoid hot spots and keep the surface of the evaporant level. If this is not done, the surface of the evaporant changes shape and the distribution of deposited film changes.

As implied earlier, the optical thickness (index times physical thickness) must be controlled accurately for each layer. There have been many schemes devised for controlling thickness, but the optical monitor and the quartz crystal microbalance are the two methods which are most widely used today.

The optical monitor measures the change in reflected or transmitted light from a substrate or glass witness slide located near the work, usually in the center of the chamber. As the thickness increases, the signal goes through maxima and minima corresponding to multiples of quarterwaves. Since the optical monitor measures optical thickness, it compensates for slight changes in index of refraction.

The quartz crystal monitor measures the change in mass of a vibrating quartz crystal as the coating builds up on it. The advantage is that the output signal varies linearly with thickness and thus it is easy to build this into a controller for both deposition rate and final thickness. The quartz crystal controller is excellent for metal films, but there is some debate over which method is best for dielectric optical thin films. Many have chosen to use the quartz crystal monitor to control the deposition cycle including rate, but use an optical monitor to control final thickness, especially for filters requiring high precision.

7.2 Photolithography

Photolithography, or optical lithography is a process used in microfabrication to selectively remove parts of a thin film or the bulk of a substrate. It uses light to transfer a geometric pattern from a photo mask to a light-sensitive chemical photoresist, or simply resist, on the substrate.

Optical lithography is a fascinating field. It requires knowledge of geometrical and wave optics, optical and mechanical systems, diffraction imaging, Fourier optics, resist systems and processing, quantification of imaging performance, and the control of imaging performance. Even the history of its development helps to stimulate new ideas and weed out less promising ones. Practitioners of optical lithography may only have a vague idea of its theory, and likewise, theoreticians may not have the opportunity to practice the technology on the manufacturing floor. In this section, we will briefly introduce the basic procedure and predict the new trend of photolithograpy.

Optical lithography has been the workhorse of semiconductor fabrication since the inception of integrated circuits. The lensless proximity printing system gradually gave way to projection-printing systems, and one-to-one replication systems became reduction systems. It took this latest form from the 0.15-NA 436-nm g-line lens, featuring resolution over 2μm with a 0.8 k_1 factor, all of the way through raising the NA until the lens became too expensive to build at that time, reducing wavelength to reposition the NA for the next round of increases, and lowering k_1 whenever the pace of NA and wavelength changes are behind the circuit shrinking roadmap, as shown in Fig.7.4.

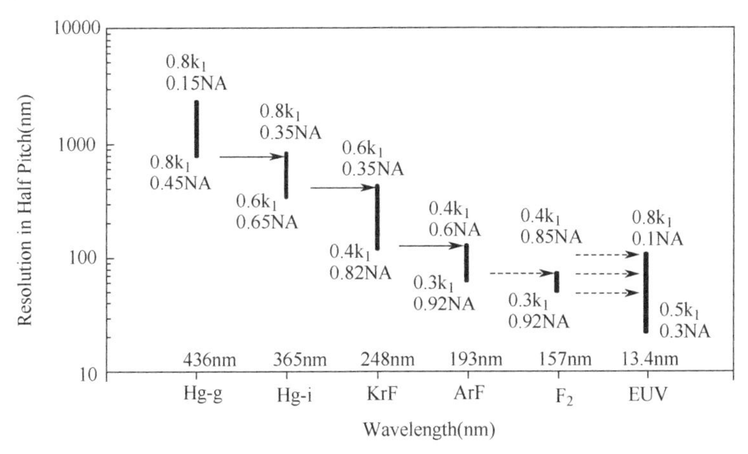

Fig.7.4 Evolution of projection optical lithography from a 0.15-NA g-line

7.2.1 Basic Procedure

A single iteration of photolithography combines several steps in sequence, as shown in Fig.7.5.

Modern cleanrooms use automated, robotic wafer track systems to coordinate the process. The procedure described here omits some advanced treatments, such as thinning agents or edge-bead removal.

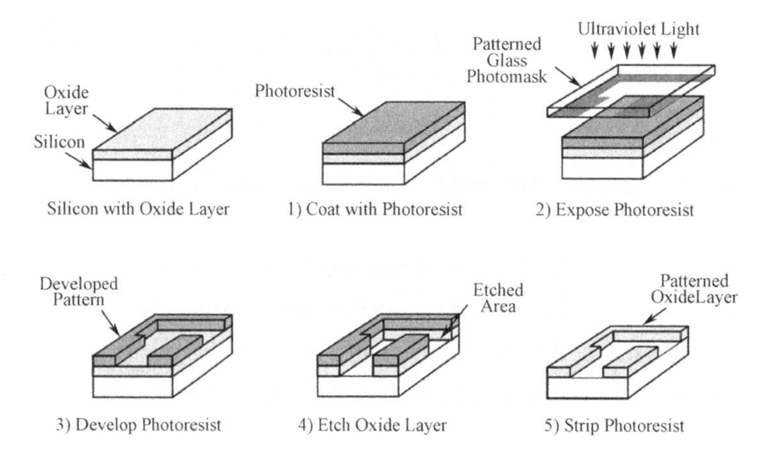

Fig.7.5 Basic photolithography process

- **Cleaning**

If organic or inorganic contaminations are present on the wafer surface, they are usually removed by wet chemical treatment.

- **Preparation**

The wafer is initially heated to a temperature sufficient to drive off any moisture that may be present on the wafer surface. Wafers that have been in storage must be chemically cleaned to remove contamination. A liquid or gaseous "adhesion promoter" is applied to promote adhesion of the photoresist to the wafer. The surface layer of silicon dioxide on the wafer reacts with HMDS (Hexamethyl Disilazane) to form tri-methylated silicon-dioxide, a highly water repellent layer not unlike the layer of wax on a car's paint. This water repellent layer prevents the aqueous developer from penetrating between the photoresist layer and the wafer's surface, thus preventing so-called lifting of small photoresist structures in the (developing) pattern.

- **Photoresist application**

The wafer is covered with photoresist by spin coating. A viscous, liquid solution of photoresist is dispensed onto the wafer, and the wafer is spun rapidly to produce a uniformly thick layer. The spin coating typically runs at 1200 to 4800 rpm for 30 to 60 seconds, and produces a layer between 0.5 and 2.5 micrometres thick. The spin coating process results in a uniform thin layer, usually with uniformity of within 5 to 10 nanometres. This uniformity can be explained by detailed fluid-mechanical modeling, which shows that the resist moves much faster at the top of the layer than at the bottom, where viscous forces bind the resist to the wafer surface. Thus, the top layer of resist is quickly ejected from the wafer's edge while the bottom layer still creeps slowly radially along the wafer. In this way, any 'bump' or 'ridge' of resist is removed, leaving a very flat layer. Final thickness is also determined by the evaporation of liquid solvents from the resist. For very small, dense features (<125 or so nm), thinner resist thicknesses (<0.5 micrometres) are needed to overcome collapse effects at high aspect ratios; typical aspect ratios are <4∶1.

The photoresist-coated wafer is then prebaked to drive off excess photoresist solvent, typically at 90 to 100℃ for 30 to 60 seconds on a hotplate.

- **Exposure and developing**

After prebaking, the photoresist is exposed to a pattern of intense light. Optical lithography typically uses ultraviolet light. Positive photoresist, the most common type, becomes soluble in the basic developer when exposed; exposed negative photoresist becomes insoluble in the (organic) developer. This chemical change allows some of the photoresist to be removed by a special solution, called "developer" by analogy with photographic developer.

- **Etching**

In etching, a liquid ("wet") or plasma ("dry") chemical agent removes the uppermost layer of the substrate in the areas that are not protected by photoresist. In semiconductor fabrication, dry etching techniques are generally used, as they can be made anisotropic, in order to avoid significant undercutting of the photoresist pattern. This is essential when the width of the features to be defined is similar to or less than the thickness of the material being etched (i.e. when the aspect ratio approaches unity). Wet etch processes are generally isotropic in nature, which is often indispensable for, microelectromechanical systems, where suspended structures must be "released" from the underlying layer. The development of low-defectivity anisotropic dry-etch process has enabled the ever-smaller features defined photolithographically in the resist to be transferred to the substrate material.

- **Photoresist removal**

After a photoresist is no longer needed, it must be removed from the substrate. This usually requires a liquid "resist stripper", which chemically alters the resist so that it no longer adheres to the substrate. Alternatively, photoresist may be removed by a plasma containing oxygen, which oxidizes it. This process is called ashing, and resembles dry etching.

7.2.2 New Trend of Photolithography

Optical lithography is marching toward the end of its legendary longevity in being a semiconductor-manufacturing workhorse. After using the 193nm wavelength and wafer immersion, polarized illumination, and improved masks, the only way to stretch optical lithography farther is to split the pitch by double patterning to reduce k_1 to 0.15 from 0.3, suggesting a half pitch of 21.5 nm. However, starting from a 32nm half pitch, there are opportunities for other technologies to compete with 193nm immersion.

We now compare the challenges of optical lithography as it faces its end and its possible contenders, which consist of EUV, MEB ML2 systems, and nanoimprint.

- **EVU lithography**

EUV lithography (EUVL) promises single-exposure, simple optical proximity correction (OPC) and has commanded worldwide development efforts at an unprecedented running cost in addition to hundreds of million dollars of sunk cost.

Will it be cost effective? The system is inevitably huge due to the need for high-speed reticule and wafer stages in vacuum. No doubt, the production-worthy hardware will be extremely expensive. The key to suppression of wafer exposure cost is to improve throughput. With resist sensitivity staying at a reasonable level of the order of 30 mJ/cm^2 to reduce line-edge roughness and shot-noise effects, the only avenue for high throughput is a high-source power, which also tends to be expensive and environmentally unfriendly, if it can even be made to work.

EUV resists may not provide cost relief. Even if the material production costs were contained, the EUV exposure tools that resist companies will need to purchase inevitably adds to the cost of EUV resists. The high cost of a defect-free EUV mask substrate is well known. The cost of a mask inspection may become a significant factor because actinic light is needed. The cost of yield loss and efforts to prevent it, due to exposure without a pellicle and radiation-induced mask contamination, has yet to be determined from field exposures at manufacture-worthy throughput.

In short, even if every technical problem in EUV lithography was solved, the high exposure cost and high raw-energy consumption may prove detrimental.

- **MEB ML2**

E-beam maskless lithography also has decades of history. It has not been the industry's workhorse for wafer imaging, except for making personalized interconnects in the earlier days at IBM. With circuit complexity ever increasing, the improvement of throughput, even though impressive by itself, could not catch up. E-beam direct write did become a workhorse for mask making.

To increase the e-beam throughput to a cost-effective level, a high degree of parallelism must be used. It can be in the number of minicolumns, the number of MEMS-fabricated microlenses, pixels in a programmable mask, or numbers of MEMS-fabricated apertures. The exploration of parallelism need not stop here. Some of the above can be favorably combined. Wafers can be exposed simultaneously in a cluster. Without clustering, throughputs on the order of 10 to 40 wph (wafer per hour) have been claimed. Some are dependent on pattern density due to space-charge-limited current. Some are not.

Even though e-beams have extremely high resolution and depth of focus, scattering in a resist spreads a well-focused beam and fundamentally limits resolution. The only way to reduce scattering is to use a smaller resist thickness. It is foreseeable that resist thickness on the order of 50nm or less must be used. A multilayer resist system or highly selective hard mask must be used.

The data rate of MEB ML2 is quite demanding. Heavy-duty data processing equipment is needed. Much of the cost in optics for an optical scanner is transferred to data-processing equipment. Fortunately, the cost reduction of data-processing equipment predictably follows Moore's law.

- **Nanoimprint lithography**

Nanoimprint, the lensless replication technique, seems to have unlimited resolution capability and the molding tool costs mere pennies. Nanoimprint has taken contact printing to a higher level. Instead of letting the mask contact the photoresist while still occupying its own space, the imaging medium is mingled with the molding template, i.e., the mask in nanoimprint. Just like Contac printing, nanoimprint is susceptible to defects. Mingling the medium with the mold does not offer promise of fewer defects.

The mold has a much shorter lifespan than the optical mask, but child molds can be replicated effortlessly using the same molding technique. Grandchildren can also be produced. Children and grandchildren must be stored and managed.

● **Comparison of the three technologies**

EUVL systems have the largest momentum and highest expectation to succeed. However, the list of challenges is still long. The higher the development cost, the more financial burden the industry will must bear, regardless of its success or failure.

MEB ML2 systems have the potential to be cost effective. They also have a long list of challenges; however, it is worth a try based on the relatively low cost to develop, the potential of becoming a workhorse, and for the absence of an increasingly difficult mask infrastructure.

Nanoimprint holds the promise of high resolution and a high height-to-width aspect ratio. Defects, template fabrication, template lifetime, throughput, and cost remain tough problems to solve.

It is not inconceivable that ArF water-immersion will be holding the last frontier for lithography, especially if double-mask exposure and high-contrast resists are realized. There may be more lithography-friendly designs, better semiconductor devices, and circuit optimization for the last frontier.

7.3 Biophotonics

Photonics is an all-encompassing light-based optical technology that is being hailed as the dominant technology for this new millennium. The invention of lasers, a concentrated source of monochromatic and highly directed light, has revolutionized photonics. Since the demonstration of the first laser in 1960, laser light has touched all aspects of our lives, from home entertainment, to high-capacity information storage, to fiber-optic telecommunications, thus opening up numerous opportunities for photonics.

A new extension of photonics is biophotonics, which involves a fusion of photonics and biology. Biophotonics deals with interaction between light and biological matter. A general introduction to biophotonics is illustrated in Fig.7.6.

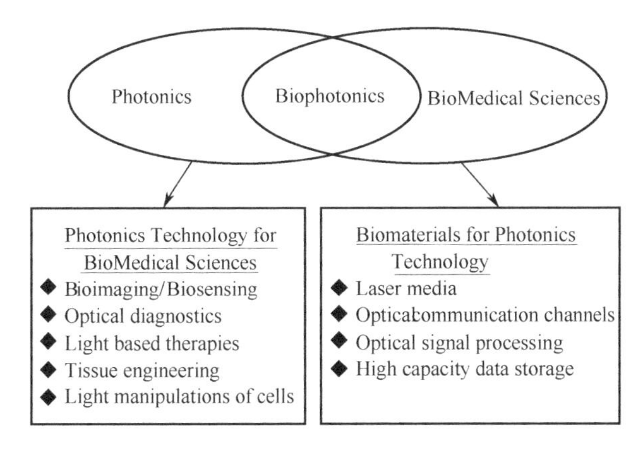

Fig.7.6 Biophotonics as defined by the fusion of photonics and biomedical sciences. The two broad aspects of biophotonics are also identified

The use of photonics for optical diagnostics, as well as for light-activated and light-guided therapy, will have a major impact on health care. This is not surprising since nature has used biophotonics as a basic principle of life from the beginning. Harnessing photons to achieve photosynthesis and conversion of photons through a series of complex steps to create vision are the best examples of biophotonics at work. Conversely, biology is also advancing photonics, since biomaterials are showing promise as new photonic media for technological applications.

As an increasingly aging world population presents unique health problems, biophotonics offers great hope for the early detection of diseases and for new modalities of light-guided and light-activated therapies. Lasers have already made a significant impact on general, plastic, and cosmetic surgeries. Two popular examples of cosmetic surgeries utilizing lasers are skin resurfacing (most commonly known as wrinkle removal) and hair removal. Laser technology also allows one to administer a burst of ultrashort laser pulses that have shown promise for use in tissue engineering. Furthermore, biophotonics may produce retinal implants for restoring vision by reverse engineering Nature's methods.

An overview of biophotonics for health care applications is presented in Fig.7.7. It illustrates the scope of biophotonics through multidisciplinary comprehensive research and development possibilities.

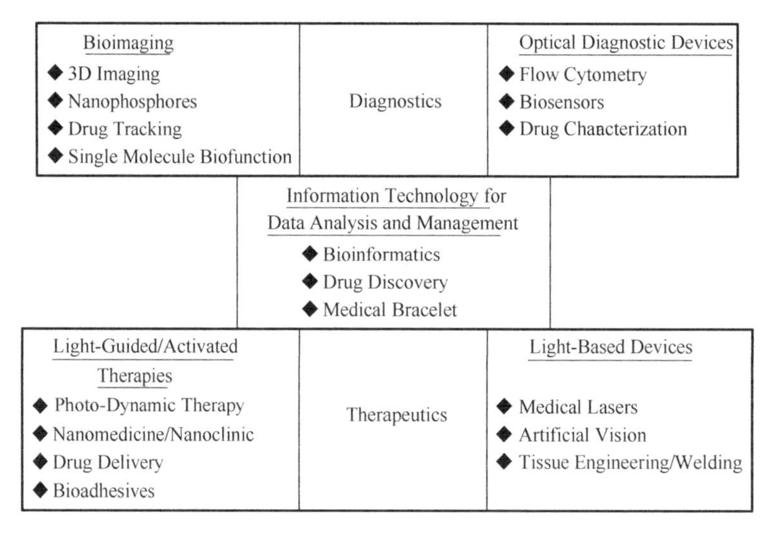

Fig.7.7 The comprehensive multidisciplinary scope of biophotonics for health care

7.3.1 Bioimaging

Biomedical imaging has become one of the most relied-upon tools in health care for diagnosis and treatment of human diseases. The evolution of medical imaging from plain radiography (radioisotope imaging), to x-ray imaging, to computer-assisted tomography (CAT) scans, to ultrasound imaging, and to magnetic resonance imaging (MRI) has led to revolutionary improvements in the quality of healthcare available today to our society. However, these techniques are largely focused on structural and anatomical imaging at the tissue or organ levels. In order to develop novel imaging techniques for early detection, screening, diagnosis, and image-guided treatment of life-threatening diseases and cancer, there is a clear need for extending imaging to the cellular and molecular biology levels. Only information at the molecular and cellular levels can lead to the detection of the early stages of the formation of a disease or cancer or early molecular changes during intervention or therapy.

The currently used medical techniques of x-ray imaging, radiography, CAT scans, ultrasound imaging, and MRI have a number of limitations. Some of these are:

- Harmful effects of ionizing radiations in the case of x-ray imaging and CAT scans
- Unsuitability of x-ray imaging for young patients and dense breasts, as well as its inability to distinguish between benign and malignant tumors
- Harmful radioactivity in radioisotope imaging
- Inability of MRI to provide specific chemical information and any dynamic information (changes occurring in real time response to a treatment or a stimulus)
- Inability of ultrasound to provide resolution smaller than millimeters as well as to distinguish between a benign and a malignant tumor

Optical imaging overcomes many of these deficiencies. Contrary to the perception based on the apparent opacity of skin, light, particularly in near-IR region, penetrates deep into the tissues. Furthermore, by using a minimally invasive endoscope fiber delivery system, one can reach many organs and tissue sites for optical imaging. Thus, one can even think of an "optical body scanner" that a physician may use some day for early detection of a cancer or an infectious disease.

Optical imaging utilizes the spatial variation in the optical properties of a biospecies, whether a cell, a tissue, an organ, or an entire live object. The optical properties can be reflection, scattering, absorption, and fluorescence. Therefore, one can monitor spatial variation of transmission, reflection, or fluorescence to obtain an optical image. The use of lasers as an intense and convenient light source to generate an optical response, whether reflection, transmission, or emission, has considerably expanded the boundaries of optical imaging, making it a most powerful technique for basic studies as well as for clinical diagnostics. Some of the benefits offered by optical imaging are:

- Not being harmful
- Imaging from size scale of 100nm to macroscopic objects
- Multidimensional imaging using transmission, reflection, and fluorescence together with spectroscopic information
- Imaging of in vitro, in vivo, and ex vivo specimens
- Information on cellular processes and tissue chemistry by spectrally resolved and dynamic imaging
- Fluorescence imaging providing many parameters to monitor for detailed chemical and dynamical information. These parameters are: Spectra, Quantum efficiency, Lifetime, Polarization
- Ability to combine optical imaging with other imaging techniques such as ultrasound
- Sensitivity and selectivity to image molecular events

The area of optical imaging is very rich, both in terms of the number of modalities and with regard to the range of its applications. It is also an area of very intense research worldwide because new methods of optical imaging, new, improved, and miniaturized instrumentations, and new applications are constantly emerging.

Fluorescence microscopy is the most widely used technique for optical bioimaging. It provides a most comprehensive and detailed probing of the structure and dynamics for in vitro, as well as in vivo, biological specimens of widely varying dimensions.

Fluorescence microscopy has emerged as a major technique for bioimaging. Fluorescence emission is dependent on specific wavelengths of excitation light, and the energy of excitation under one photon absorption is greater than the energy of emission (the wavelength of excitation light is shorter than the wavelength of emission light). Fluorescence has the advantage of providing a very high signal-to-noise ratio, which enables us to distinguish spatial distributions of even low concentration species. To utilize fluorescence, one can use endogenous fluorescence (autofluorescence) or one may label the specimen (a cell, a tissue, or a gel) with a suitable molecule (a fluorophore, also called fluorochrome) whose distribution will become evident after illumination. The fluorescence microscope is ideally suited for the detection of particular fluorochromes in cells and tissues.

The fluorescence microscope that is in wide use today follows the basic "epifluorescence excitation" design utilizing filters and a dichroic beam splitter. The object is illuminated with fluorescence excitation light through the same objective lens that collects the fluorescence signal for imaging. A beam splitter, which transmits or reflects light depending on its wavelength, is used to separate the excitation light from the fluorescence light. In the arrangement, shown in Fig.7.8, the shorter-wavelength excitation light is reflected while the longer-wavelength emitted light is transmitted by the splitter.

With the advent of different fluorochromes/ fluorophores, specifically targeting different parts of the cells or probing different ion channel processes (e.g., Ca^{2++} indicators), the fluorescence microscopy has had a major impact in biology. The development of confocal microscopy has significantly expanded the scope of fluorescence microscopy.

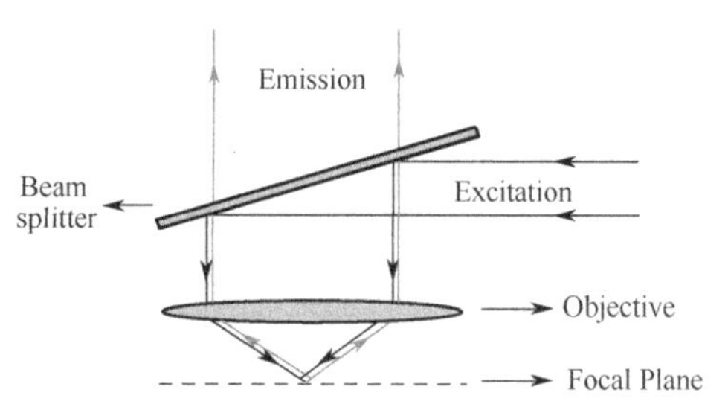

Fig.7.8 Basic principle of epi-fluorescence illumination

7.3.2 Optical Biosensors

The field of biosensors has emerged as a topic of great interest because of the great need in medical diagnostics and, more recently, the worldwide concern of the threat of chemical and bioterrorism. The constant health danger posed by new strands of microbial organisms and spread of infectious diseases is another concern requiring biosensing for detecting and identifying them rapidly. Optical biosensors utilize optical techniques to detect and identify chemical or biological species. They offer a number of advantages such as the ability for principally remote sensing with high selectivity and specificity and the ability to use unique biorecognition schemes.

Biosensors are analytical devices that can detect chemical or biological species or a microorganism. They can be used to monitor the changes in vivo concentrations of an endogenous specie as a function of a physiological change induced internally or by invasion of a microbe. Of even more recent interest is the use of biosensors to detect toxins, bacteria, and viruses because of the danger posed by chemical and biological terrorism. Biosensors thus find a wide range of applications, such as Clinical diagnostics, Drug development, Environmental monitoring (air, water, and soil), and Food quality control, etc..

A biosensor in general utilizes a biological recognition element that senses the presence of an analyte (the specie to be detected) and creates a physical or chemical response that is converted by a transducer to a signal. The general function of a biosensor system is described in Fig.7.9.The sampling unit introduces an analyte into the detector and can be as simple as a circulator. The recognition element binds or reacts with a specific analyte, providing biodetection specificity. Enzymes, antibodies or even cells such as yeast or bacteria have been used as biorecognition elements. Stimulation, in general, can be provided by optical, electric, or other kinds of force fields that extract a response as a result of biorecognition. The transduction process transforms the physical or chemical response of biorecognition, in the presence of an external stimulation, into an optical or electrical signal that is then detected by the detection unit. The detection unit may include pattern recognition for identification of the analyte. In the most commonly used form of an optical biosensor, the stimulation is in the form of an optical input. The transduction process induces a change in the phase, amplitude, polarization, or frequency of the input light in response to the physical or chemical change produced by the biorecognition process. Some of the other approaches use electrical stimulation to produce optical transduction (e.g., an electroluminescent sensor) or an optical stimulation to produce electrical transduction (e.g., a photovoltaic sensor).

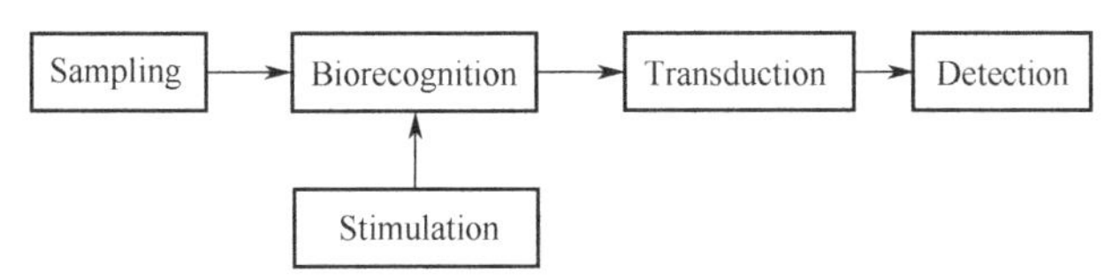

Fig.7.9 General scheme for biosensing

The field of biosensors has been active over many decades. The earlier successes were sensors utilizing electrochemical response (Janata, 1989). This type of sensor still tends to dominate the current commercial market. However, progress in fiber optics and integrated optics (such as channel waveguides and surface plasmon waves) and the availability of microlasers (solid-state diode lasers) have made optical biosensors a very attractive alternative for many applications. An optical biosensor, in general, utilizes a change in the amplitude (intensity), phase, frequency or polarization of light created by a recognition element in response to a physiological change or the presence of a chemical or a biologic (e.g., microorganism). Enhancement of the sensitivity and selectivity of the optical response is achieved by immobilizing the biorecognition element (such as an antibody or an enzyme) on an optical element such as a fiber, a channel waveguide, or a surface plasmon propagation where light confinement produces a strong internal field or an evanescent (exponentially decaying) external field. Thus, the main components of an optical biosensor include

(i) a light source, (ii) an optical transmission medium (fiber, waveguide, etc.), (iii) immobilized biological recognition element (enzymes, antibodies or microbes), (iv) optical probes (such as a fluorescent marker) for transduction, and (v) an optical detection system.

Some of the advantages offered by an optical biosensor are:

- Selectivity and specificity
- Remote sensing
- Isolation from electromagnetic interference
- Fast, real-time measurements
- Multiple channels/multiparameters detection
- Compact design
- Minimally invasive for in vivo measurements
- Choice of optical components for biocompatibility
- Detailed chemical information on analytes

Wadkins et al. (1998) used a scheme, shown in Fig.7.10, that used patterned antibody channels.

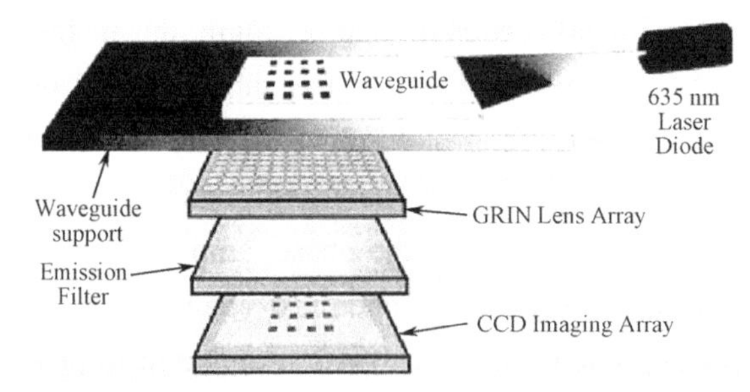

Figure 7.10 Array biosensors developed by Ligler, Golden, and co-workers at the Naval Research Laboratory

They demonstrated the detection of Y. pestis F1 (a Gram-negative rod-shaped bacterium) in clinical fluids such as whole blood, plasma, urine, saliva, and nasal secretion.

7.3.3 Microarray Technology for Genomics and Proteomics

Microarray technology provides a powerful tool for high-throughput rapid analysis of a large number of samples. This capability has been of significant value in advancing the fields of Genomics, Proteomics, and Bioinformatics, which are at the forefront of modern structural biology, molecular profiling of diseases, and drug discovery. Biophotonics has played an important role in the development of microarray technology, since optical methods are used for detection and readout of microarrays.

Microarray technology is a powerful, universally applicable analytical tool for rapid and simultaneous detection and analysis of a large number of biological assemblies with high sensitivity and precision. Thousands of DNA and RNA species can be simultaneously probed using a DNA microarray to provide detailed insight into cellular phenotyping and genotyping. Phenotyping is the process of identifying cells with specific markers. When antibodies are used, the process is called immunophenotyping. When genetic constructs are used, it is called genotyping.

The various types of biological microarrays are compiled in Fig.7.11. Of these the DNA microarray technology is most developed and in wide usage all around the world. The term gene chip or biochip is also often used to describe a DNA microarray because the approaches used to fabricate DNA microarrays often involve processing analogous to that used to produce semiconductor microchips. Since its commercial availability in 1996, DNA microarrays have now become a major tool for genomics and drug discovery.

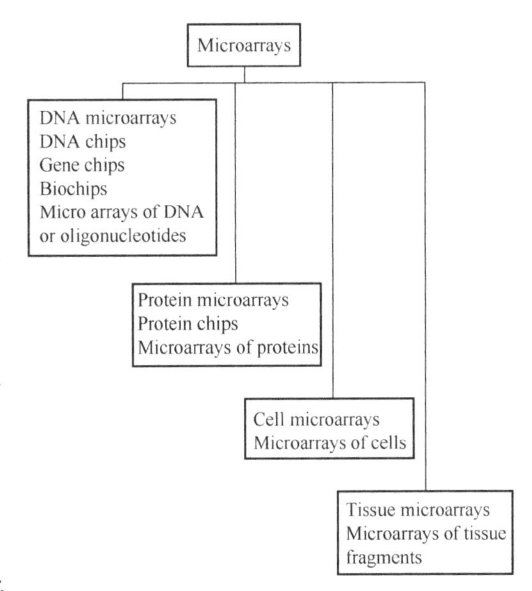

Fig.7.11 Various Biological Microarrays

Protein microarrays or protein chips utilize microarrays of immobilized fusion proteins (proteins that fuse with other proteins) or antibodies. However, the protein chips currently are not sufficiently robust for high throughput studies.

Cell microarrays are a relatively new development that utilizes live cells expressing a c-DNA of interest.

Tissue microarrays developed in the laboratory of Kallioniemi provides a new high-throughput tool for the study of gene dosage and protein expression patterns in a large number of individual tissues. This tool can provide a rapid and comprehensive molecular profiling of cancer and other diseases without exhausting limited tissue resources.

The two main pieces of hardware for microarray technology are (i) the microarray slide spotter and (ii) the microarray scanner. The two main approaches used to fabricate DNA microarrays are described in Fig.7.12.

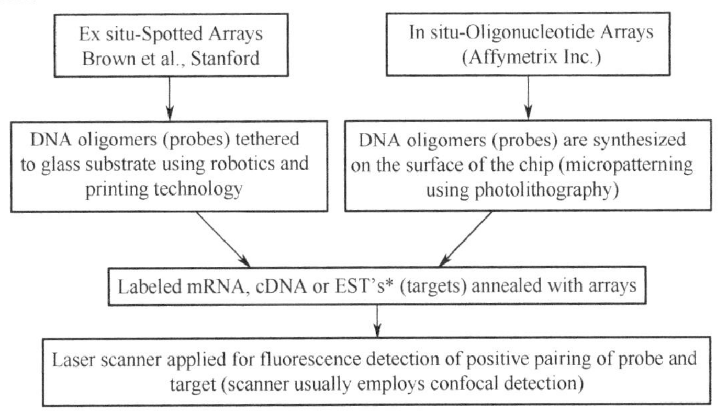

Fig.7.12 Two Main Approaches for DNA Microarray Technology

7.3.4 Light-Activated Therapy: Photodynamic Therapy

An important area of biophotonics is the use of light for therapy and treatment. The use of light to activate a photosensitizer eventually leads to the destruction of cancer or a diseased cell. This procedure is called photodynamic therapy (abbreviated as PDT) and constitutes a multidisciplinary area that has witnessed considerable growth in activities worldwide.

Photodynamic therapy (PDT) has emerged as a promising treatment of cancer and other diseases utilizing activation of an external chemical agent, called a photosensitizer or PDT drug, by light. This drug is administered either intravenously or topically to the malignant site as in the case of certain skin cancers. Then light of a specific wavelength, which can be absorbed by the PDT photosensitizer, is applied. The PDT drug absorbs this light, producing reactive oxygen species that can destroy the tumor.

The key steps involved in photodynamic therapy are shown in Fig.7.13.They are:

- Administration of the PDT drug
- Selective longer retention of the PDT drug by the malignant tissue
- Delivery of light, generally laser light, to the malignant tissue site
- Light absorption by the PDT drug to produce highly reactive oxygen species that destroy cancer cells with minimal damage to surrounding healthy cells
- Clearing of the drug after PDT to reduce sunlight sensitivity

As indicated above, PDT relies on the greater affinity of the PDT drug for malignant cells. When a PDT drug is administered, both normal and malignant cells absorb the drug. However, after a certain waiting period ranging from hours to days, the concentration of the PDT drug in the normal cell is significantly reduced. Recent studies with tumor-targeting agents attached to the PDT drug have shown that their waiting period can be reduced to a matter of a few hours. In contrast, the malignant cells still retain this drug, thus producing a selective localization of this drug in the malignant tissue site. At this stage, light of an appropriate wavelength is applied to activate the PDT drug, which then leads to selective destruction of the malignant tissue by a photochemical mechanism (nonthermal, thus no significant local heating). In the case of cancer in an internal organ such as a lung, light is administered using a minimally invasive approach involving a flexible fiber-optic delivery system. In the case of a superficial skin cancer, a direct illumination method can endoscopic be used. Since coherence property of light is not required, any light source such as a lamp or a laser beam can be used. However, to achieve the desired power density at the required wavelength, a laser beam is often used as a convenient source for this treatment. The use of a laser beam also facilitates fiber-optic delivery.

There are only two currently approved drugs: photofrin and verteroporfin. In all cases, photofrin has been used as a photosensitizer that is activated at 630nm (see Fig.7.14). Fluorescence images show a tumor on the back before and after irradiation with PDT light. Drug dose: 5mg/kg b.w.; uptake time: 12hr; excitation light: 660nm. This PDT drug is administered by an intravenous.

7.3.5 Nanotechnology for Biophotonics: Bionanophotonics

Some describe us as living in an era of Nanomania where there is a general euphoria about nanoscale science and technology. The fusion of nanoscience and nanotechnology with biomedical research has also broadly impacted biotechnology. Imagine nanosubmarines navigating through our bloodstreams and destroying nasty viruses and bacteria. Imagine nanorobots hunting for cancer cells throughout our body, finding them, then reprogramming or destroying them. A subject of science fiction at one time has now been transformed into a future vision showing promise to materialize.

The fusion of nanoscience and nanotechnology into biomedical research has brought in a true revolution that is broadly impacting biotechnology. New terms such as nanobioscience, nanobiotechnology, and nanomedicine have come into existence and gained wide acceptance.

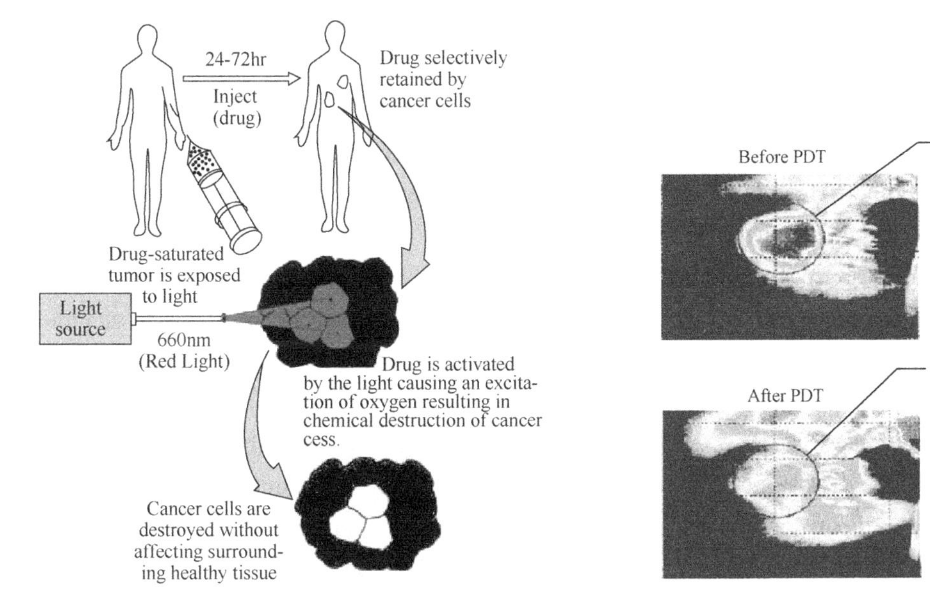

Fig.7.13 The steps of photodynamic therapy with a specific PDT drug

Fig.7.14 Fluorescence images showing the effectiveness of PDT treatment of a tumor

Table 7.1 lists some nanotechnology frontiers in bioscience. Nanophotonics is an emerging field that deals with optical interactions on a scale much smaller than the wavelength of light used (Shen et al., 2000). The three major areas of nanophotonics are shown in Fig.7.15.

Table 7.1 Nanotechnology frontiers in bioscience

Biophysics	**Structural Biology**
Nanomechanics	Protein folding, design
Optical traps	"Rational"drug/ligand design
Flexible-probe methods	Novel and improved methods
Single-molecule methods	**Computational Biology**
FRET	Protein folding, design
New labels	"Rational"drug/ligand design
New reporters	Bioinformatic design, regulation
New imaging, microscopies	**Biotronics—Biomolecules on Chips**
Scanned-probe	DNA and protein nanotrays
Combinations	Sensors, detectors, diagnostics
	Labs-on-a-chip
Biochemistry	**Biofabrication**
Single-molecule enzymology	Nanoparticle delivery systems
Single-molecule kinetics	Biomaterials, tissue engineering
Single-molecule sequencing	Implants, prosthetics

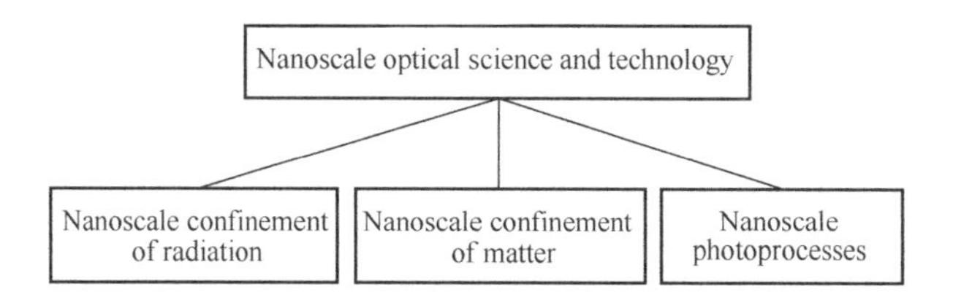

Fig.7.15 Three major areas of nanophotonics

Nanoscale confinement of radiation is achieved in a near-field geometry. This allows one to break diffraction barriers and obtain optical resolution to less than 100nm.Near-field microscopy is becoming a powerful biomedical research tool to probe structure and functions of submicron dimension biological species such as bacteria. Nanoscale confinement of matter is achieved by producing nanoparticles, nanomers, nanodomains, and nanocomposites. The nanosize manipulation of molecular architecture and morphology provides a powerful approach to control the electronic and optical properties of a material. An example is a semiconductor quantum dot, a nanoparticle whose electronic band gap and thus the emission wavelength are strongly dependent on its size. Nanoscale control of the local structure in a nanocomposite, consisting of many domains and separated only on the nanometer scale, provides an opportunity to manipulate excitedstate dynamics and electronic energy transfer from one domain to another. Such nanostructured materials can provide significant benefits in fluorescence resonance energy transfer (FRET) imaging and in flow cytometry respectively.

Nanoscale photoprocesses such as photopolymerization provide opportunities for nanoscale photofabrication. Near-field lithography can be used to produce nanoarrays for DNA or protein detection. The advantage over the microarray technology is the higher density of arrays obtainable using near-field lithography, thus allowing one to use small quantities of samples. This is a tremendous benefit for protein analysis in the case when the amount of protein produced is very minute and there is no equivalent of DNA PCR amplification for proteins to enhance the detection.

The applications of nanophotonics to biomedical research and biotechnology range from biosensing, to optical diagnostics, to light activated therapy. Nanoparticles provide a highly useful platform for intracellular optical diagnostics and targeted therapy. The area of usage of nanoparticles for drug delivery has seen considerable growth.

Levy has developed the concept of a nanoclinic, a complex surface functionalized silica nanoshell containing various probes for diagnostics and drugs for targeted delivery. An illustrated representation of a nanoclinic is shown in Fig.7.16.

Nanoclinics provide a new dimension to targeted diagnostics and therapy. These nanoclinics are produced by multistep nanochemistry in a reverse micelle nanoreactor.

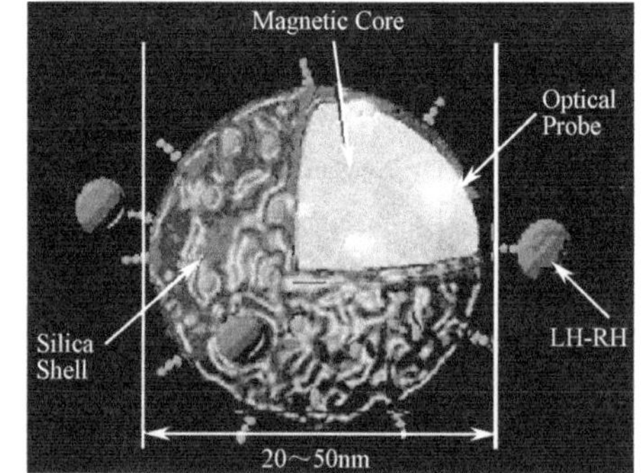

Fig.7.16 Illustrated representation of a nanoclinic

7.4 3D Display Technology

In "3D", "D" is the first letter of "dimension", and "3D" means three-dimensional space. Comparing with ordinary 2D images, 3D technology can make the picture become stereo and lifelike. Therefore images are not confined to flat of the screen, as if walking outside of the screen, making the audiences in them. It's plain fact that 3D is everywhere these days from movies and games to laptops and handhelds.

7.4.1 Classification of 3D Display Technology

Close just your right eye. Now just your left. Now your right. It's like in Wayne's World: camera one, camera two. You must have noticed that things change position a bit. This, of course, is because your eyes are a few inches apart; this is called the "interocular distance" and it varies from person to person. Note also that when you look at something close, objects appear in double in the background. Look at the corner of the screen. You can see the chair and window back there are doubled because you're actually rotating your eyes so they both point directly at what you're focusing on. This is called "convergence," as Fig.7.17 shows, and it creates a sort of X, the center of the X being what's being focused on.

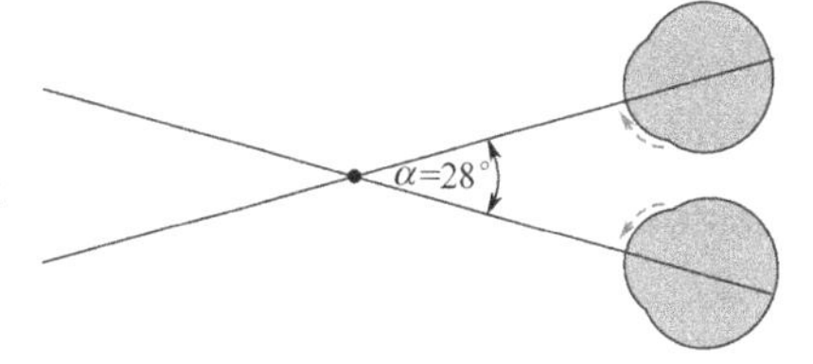

Fig.7.17 Interocular distance

So the objects at the center of the X are aligned at the same points on your respective retinas, but because of the interocular, that means that things in front and behind of that X are going to hit different points on those retinas, resulting in a double image. You can see through the double image because what's blocked for one eye is seen by the other, though from a slightly different point of view.

Next we have focus. Your eye focuses differently from most cameras, but the end result is similar. Try holding your finger out at arm's length and focusing on it, then on something distant that is directly behind it. Obviously the focus changes and you may have noticed that while the distant object was blurry while you looked at your finger, your finger (in double) was in focus while you looked at the distant object. That has to do with the optic qualities of your eye, which is important here. More important, however, is this: try it again, and this time pays attention to the feeling in your eye. You can feel that sort of like something's moving but you can't tell exactly what. Your eyes are rotating a bit, but only a few degrees, in order to converge further out, but more importantly, you're feeling the muscles of your eye actually crush or stretch out the lens of the eye, changing the path and internal focal plane of the light entering your eye. Farsightedness and nearsightedness occur at this stage, when either the lens or the eyeball itself is misshaped, resulting in focus being skewed or difficult to resolve one way or another.

Once the two images have been presented to your retinas, they pass back through the optic nerve to various visual systems, where an incredibly robust real-time analysis of the raw data is performed by several areas of the brain at once. Some areas look for straight lines, some for motion, some perform shortcut operations based on experience and tell you that yes indeed, the person did go behind that wall, they did not disappear into the wall, and that sort of thing. Eventually (within perhaps 50 milliseconds) all this information filters up into your consciousness and you are aware of color, depth, movement, patterns, and distinct objects within your field of view, informed mainly by the differences between the images hitting each of your retinas. It's important to note that vision is a learned process, and these areas in your visual cortex are "programmed" by experience as much as by anatomy and, for lack of a better term, instinct.

Although 3D display technologies are various, the most basic image-forming principle is similar. Fig.7.18 shows that these technologies use the human left and right eyes to receive different frames respectively, then with superposing and reconstructing image information, the brain form a fore-back, up-down, left-right, far-near image which is three-dimensional.

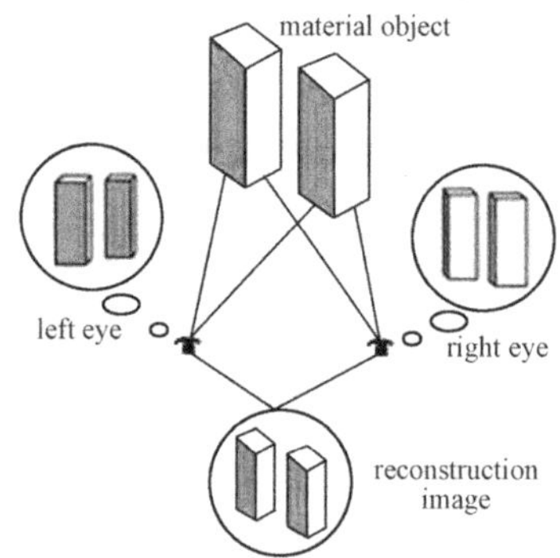

Fig.7.18 The basic image-forming principle demonstration

3D display technology can be divided into two categories: aided-viewing and free viewing and more careful classification can be visible in table 7.2.

Tab.7.2 Classification of 3D display techniques

Eye-addressing method	**Multiplex method**	**Origin of waves**	**Number of views**	**Motion parallax**
Aided-viewing	Colour	Fixed image plane	Two	Optional(for a single head-tracked viewer)
	Polarization	Eye-gaze-related		
	Time	Image plane		
	Location	Pixel-specific location		
Free-viewing (auto stereoscopic)	Direction (e.g.by diffraction, refraction, reflection, occlusion)	Fixed image plane	Two	Original(for a single head-tracked viewer)
		Eye-gaze-related image plane	Multiple	Inherent(for a small group of viewers)
	Volumetric display	Multiple fixed image planes	Unlimited	Inherent(for a small group of viewers)
	Electroholography	Entire space		

Free viewing 3D display technology is now used mainly at public business occasions, and it will be applied to mobile phones and other portable devices. While in the field of family consumption, whether displays, projector or TV, it needs to cooperate with 3D glasses.

7.4.2 Aided-viewing

The aided-viewing 3D display technology includes three main types: anaglyphic 3D, polarization 3D and active shutter 3D. They are also usually called the method of colour division, light division and time division.

- **Anaglyphic 3D**

Anaglyphic 3D display technology need to use red-blue (or red-green) color 3D filter glasses to aid. The history of this technology is the longest, and the principle of image-forming is simplest. It only takes a few dollars, while its 3D display effect is the worst.

Firstly, anaglyphic 3D gives light spectrum by the rotating filter wheel. Then, it uses optical filter with different color to filter the light, making a picture produce two images. Thus the person's each eye can see different images. However, this method will produce partial color picture edge.

- **Polarization 3D**

Polarization 3D is used together with the passive polarized glasses, thus the request of the display equipment's brightness is higher. While polarization 3D display technology has better imaging effect comparing to anaglyphic 3D display technology, its cost is not too much as well. The technology is used widely in most cinemas at present.

Polarization 3D uses the principle of vibration direction to disintegrate the original image. Firstly it divides images into vertical polarized light and level polarized light two groups images, then the left and right 3D glasses adopt polarized lenses with different polarized direction respectively. As Fig.7.19 shows the course redrawing stereo images with polarization 3D glasses, people's eyes can receive two groups of pictures, and then combine stereo images together through brains.

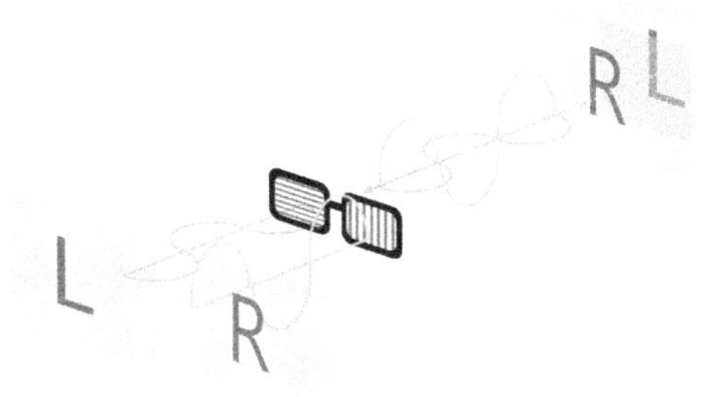

Fig.7.19 The course of redrawing stereo images with polarization 3D glasses

We can see that when stacking the same direction lens, euphotic rate decreases obviously; while stacking the vertical direction lens, it is far from to transmit light in polarization 3D glasses.

- **Active shutter 3D**

Active shutter 3D display technology is widely applied in television and the projector. It has many resources relatively, and its image effect is outstanding. Though the matching 3D glasses cost high expenses, it is highly praised by a lot of manufactures.

Active shutter 3D achieves 3D effect mainly through improving the refresh rate of pictures. It puts the image in two by frame, forms two pictures corresponding to the left and right eyes, and interlaces to display continuously. At the same time the infrared signal transmitters will control the left and right lens switch of active shutter 3D glasses synchronously, making the left and right eyes see the corresponding picture at the right moment. This technology can keep the original picture resolution, make users enjoy the real full high definition (HD) 3D effect very easily, and won't reduce screen brightness.

7.4.3 Free Viewing

With practice, most readers can view stereo pairs without the aid of blocking devices using a technique called free viewing. There are two types of free viewing, distinguished by how the left and right eye images are arranged. In parallel or uncrossed viewing, the left eye image is to the left of the right eye image. In transverse or cross viewing, they are reversed and crossing the eyes to form an image in the center is required. Some people can do both types of viewing, some only one, some neither. In Fig.7.20, the eye views have been arranged in left/right/left order. To parallel view, look at the left two images. To cross view, look at the right two images.

- **Parallax barrier displays**

A parallax barrier consists of a series of fine vertical slits in an otherwise opaque medium. The barrier is positioned close to an image that has been recorded in vertical slits and back lit. If the vertical slits in the image have been sampled with the correct frequency relative to the slits in the parallax barrier, and the viewer is the required distance from the barrier, then the barrier will occlude the appropriate image slits to the right and left eyes respectively and the viewer will perceive an autostereoscopic image. The images can be made panoramic to some extent by recording multiple views of a scene. As the viewer changes position, different views of the scene will be directed by the barrier to the visual system. The number of views is limited by the optics and, hence, moving horizontally beyond a certain point will produce "image flipping" or a cycling of the different views of the scene.

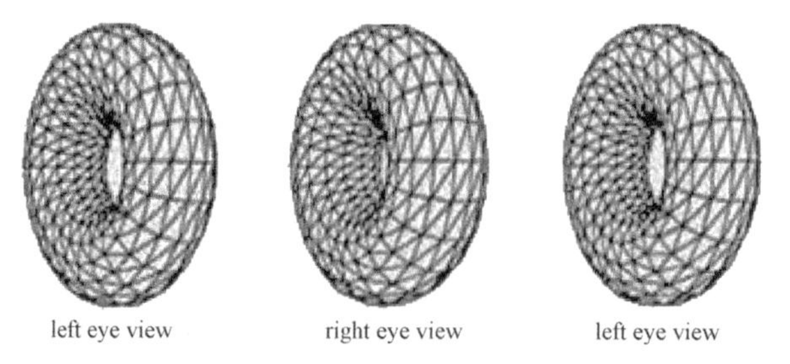

Fig.7.20 Free viewing examples

High resolution laser printing has made it possible to produce very high quality images: the barrier is printed on one side of a transparent medium and the image on the other. This technique was pioneered in the early 1990's to produce hard copy displays.

- **Lenticular lens**

Fig.7.21 shows the principle of lenticular lens display. A lenticular sheet consists of a series of semi-cylindrical vertical lenses called "lenticles," typically made of plastic. The sheet is designed so the parallel light entering the front of the sheet will be focused onto strips on the flat rear surface. By recording an image in strips consistent with the optics of the lenticles as in the case of the parallax barrier display, an autostereoscopic panoramic image can be produced. Because the displays depend on refraction vs. occlusion the brightness of

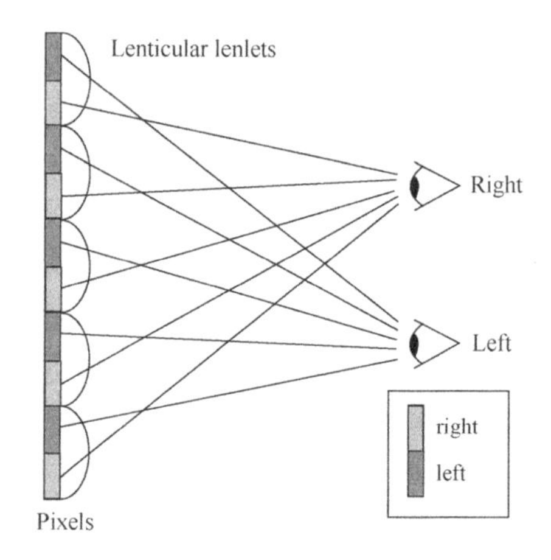

Fig.7.21 The principle of lenticular lens display

a lenticular sheet display is usually superior to the parallax barrier and requires no back-lighting. Such displays have been mass produced for many years for such hardcopy media as postcards.

In these two techniques the image is recorded in strips behind the parallax barrier or the lenticular sheet. Although the techniques are old, recent advances in printing and optics have increased their popularity for both hardcopy and auto stereoscopic display devices. In both the lenticular and parallax barrier cases, multiple views of a scene can be included providing motion parallax as the viewer moves his/her head from side to side creating what is called a panoramagram. Recently, parallax barrier liquid-crystal imaging devices have been developed that can be driven by a microprocessor and used to view stereo pairs in real time without glasses.

- **Holographic stereogram**

Most readers are familiar with holographic displays, which reconstruct solid images. Normally a holographic image of a three dimensional scene will have the "look around" property. A popular combination of holography and stereo pair technology, called a holographic stereogram, involves recording a set of 2D images, often perspective views of a scene, on a piece of holographic film. The film can be bent to form a cylinder so the user can walk around the cylinder to view the scene from any aspect. At any point the left eye will see one view of the scene and the right eye another or the user is viewing a stereo pair.

Conventional display holography has long been hampered by many constraints such as limitations with regard to color, view angle, subject matter limitations, and final image size. Even with the proliferation of holographic stereogram techniques in the 1980s, the majority of the constraints remained. Zebra Imaging, Inc. expanded on the developments in one-step holographic stereogram printing techniques and has developed the technology to print digital full-color reflection holographic stereograms with a very wide view angle (up to 110°), unlimited in size, and with full parallax.

Zebra Imaging's holographic stereogram technique is based on creating an array of small (1mm or 2mm) square elemental holographic elements. Much like the pixels of two-dimensional digital images, hogel (short for holographic pixel) arrays can be used to form complete images of any size and resolution. Each hogel is a reflection holographic recording on pan-chromatic photopolymer film. The image recorded in each hogel is of a two-dimensional digital image on a spatial light modulator (SLM) illuminated with laser light in the three primary colors: red, green, and blue.

7.5 Infrared Detection Technology

Infrared (IR) is an electromagnetic spectrum at a wavelength that is longer than visible light. It can not be seen but it can be detected. Objects that generate heat also generate infrared radiation and those objects include animals and the human body whose radiation is strongest at a wavelength of 9.4 μm. Infrared in this range will not pass through many types of material that pass visible light such as ordinary window glass and plastic. However, it can pass through, with some attenuation, material that is opaque to visible light such as germanium and silicon.

The infrared detection region of the electromagnetic spectrum ranges from 0.78 to 1000 microns (μm) and can be further subdivided into the near, mid and far IR regions which range from 0.78 μm to 2.5μm, 2.5μm to 50μm and 50μm to 1000μm, respectively. The most common wavelengths for detection applications occur in the mid IR region and range between 2.5μm and 15μm.

A blackbody is a theoretical construct that absorbs all the radiant energy striking it and radiates energy at the maximum possible rate per unit area at each wavelength for any given temperature.

In quantum mechanics, the dependence of the radiation density on its frequency and temperature is given by Planck radiation formula, i.e.,

$$R_0 = \frac{2\pi f^3 h}{c^2 \left\{ \exp\left(\frac{hf}{kT} \right) - 1 \right\}} \tag{7.8}$$

where h=6.62×10^{-34}J • s is Planck constant; c=3×10^5km/s is the velocity of electromagnetic waves in vacuum; k =1.38×10^{-23}J/K is the Boltzmann constant; T is the absolute temperature of the radiating body; f is frequency; R_0 is the spectral density of radiation equal to power radiation on frequency f in the band of 1.0 Hz by 1.0 m^2 of the radiator.

The frequency f_m on which the spectral density of radiation has the maximum is determined by Wien formula:

$$f_m = 1.03 \times 10^5 \cdot T \tag{7.9}$$

At present, there are several different detection techniques that utilize IR spectroscopy. They include active infrared detection technology, passive IR detection technology and thermal infrared remote sensing technology etc..

7.5.1 Active Infrared Detection Technology

An active infrared detector includes a radiation source and an infrared sensor which is sensitive to interruptions in the radiation sensed from the source. These detectors are used as intrusion detectors by providing a path of radiation from the source to the sensor in a place where the path is likely to be interrupted by an intruder.

The proposed active infrared method of motion detection has the advantage of fast response speed of a relatively large sensor. This advantage permits simpler optical system design, especially for wide fields of view. Besides, it is insensitivity to mechanical and acoustic noise, which presents substantial problems in the passive infrared (PIR) sensors. Low production cost is another advantage of these active infrared detectors.

In active infrared systems, there are two-piece elements which are consisting of an infrared transmitter and an infrared receiver. And there is a 3/8 inch infrared beam between the transmitter which is placed on one side of the trail and the receiver which is placed on the other side of the trail. The transmitter and the receiver can be separated whose distance is as much as 150 feet.

The transmitter emits a beam of light into the scan zone. The light, which is reflected by the background returns to the receiver, which constantly monitors the scan zone. The infrared light is interrupted when a person or object enters the zone. It then sends a signal to the controller system, which is wired into the door controls. One variation of this operating mode is called 'background suppression'. This is when the receiver only detects a change in the reflected light when a person or object enters the scan zone thus causing a reflectance variation of the light, sending a signal to the microcontroller thus trigger the alarm of the security system.

House security system is one of security that truly related to burglar or safety alarm system. Burglar and safety alarms are found in electronic form nowadays. Sensors are connected to a control unit via either a low-voltage hardwire which in turn connects to a means for announcing the alarm to elicit response.

In a new construction systems are predominately hardwired for economy while in retrofits wireless systems may be more economical and certainly quicker to install. Some systems are dedicated to one mission: handle fire, intrusion, and safety alarms simultaneously.

In common security system, the lights are triggered by motion and give the impression to user that someone is at home and able to see the burglar. Infrared motion detectors placed in house security system in crucial areas of the house can detect any burglars and alert the home owner or police.

The first security system invented, house alarms are triggered by the release of a pressure button fitted into a door or window frame. This basic alarm is fundamentally flawed as the entire intruder need to silence the alarm to close the door or window.

While various systems on the market ranging from inexpensive house security alarms to highly sophisticated systems require the professional installation. All modern alarms are based on the same foundation, the electric circuit which is completed either when the door is opened or closed depending on the security system designed.

The alarm is triggered when the circuit is altered and will not be silenced until a code is punched into the control panel. The most expensive and complicated alarm systems might also involve a combination of motion sensors and pressure pads to ensure even the most cunning intruder doesn't get his hands on treasures.

7.5.2 Passive Infrared Detection Technology

PIR motion detectors always use thermal sensors, detecting the small temperature increase when the sensor element is exposed to radiation and absorbs it. Quantum detectors are not practical for PIR due to their need for cooling.

We will see that thermal radiation is represented by Plank's law. The corresponding theory was developed from a theoretical physical model of a radiator: the blackbody, while the radiation from real bodies is derived through a coefficient called emissivity. Although in general thermal radiation is emitted over the whole spectrum, its distribution is not uniform. It is found to be concentrated over a relatively narrow region around a wavelength denoted by λ_{peak}. Around 70% of the energy emitted is found to be concentrated between $0.5\lambda_{peak}$ and $2\lambda_{peak}$. This wavelength is derived from the blackbody temperature by Wien's law, i.e.,

$$\lambda_{peak} T = 2898 \tag{7.10}$$

where λ_{peak} is expressed in micro and T Kelvins. Thus, for objects whose temperatures are around ambient (300K) we derive that λ_{peak}=10μm. It is therefore the spectral region between 5 and 20μm which is the most suitable for passive observation of the surrounding scene.

On the other hand, we know that the photon energy of monochromatic radiation having a wavelength by

$$E = hc/\lambda = 1.241/\lambda \tag{7.11}$$

where h is Planck's constant and c is the speed of light in the propagation medium; for λ in micrometers, E is given in electron-volt. From it we can immediately derive that at the wavelength being considered, E=0.12eV. Moreover, since the photon energy varies as an inverse function of wavelength, we note that the infrared radiation will be more difficult to detect than visible radiation, and even harder to detect than the ultraviolet. Figure 7.22 shows the spectrum of optical radiation.

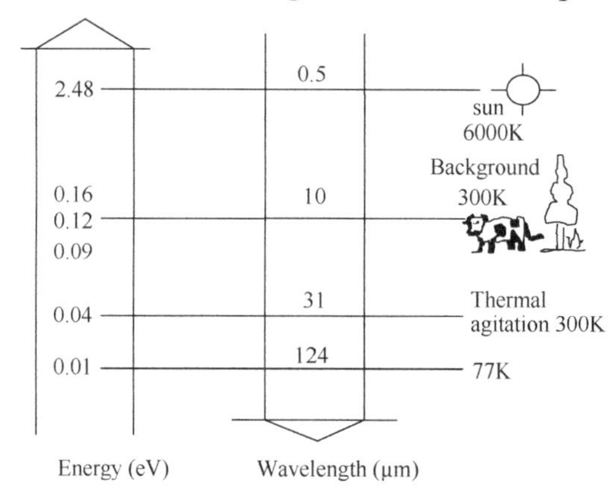

Fig.7.22 Spectrum of optical radiation

It is usually the case that the radiation arrives at the detector after having crossed the atmosphere. Although photons in the infrared have a low energy by comparison with visible radiation, their energy remains sufficient after interacting with the molecules which make up the gaseous layer. As opposed to radar waves or radio frequencies, the extinction of radiation is therefore significant.

It can even be total at ground level in certain absorption bands as soon as the propagation distance exceeds several hundred meters. Only a few spectral regions, called atmospheric window, have a transparency sufficient to allow an optical coupling over distances greater than several kilometers; they are therefore of great practical interest. In regard to radiation sources they are characterized in the following way:

- window I for $\lambda \leqslant 2.8$μm, radiation from very hot sources and solar reflections;
- window II for $3 \leqslant \lambda \leqslant 5$μm, emission of hot bodies and combustion gases;
- window III for $8 \leqslant \lambda \leqslant 14$μm, emission of hot bodies with temperatures near ambient;

In all cases the conditions for atmospheric propagation largely depend on the geographical characteristics and the local meteorological conditions.

In 1981, Marcel Züblin picked up the idea of switching light with a PIR. The idea was not new, but he postulated 180 field of view, 2-wire operation and an affordable price as the imperative ingredients for the product to replace the manual wall switch, and he was determined to make it work.

More recently, in 1992, PIR motion detectors were not only considered to switch light for comfort and safety, but to control energy in buildings, mainly to switch heating, ventilation, air-conditioning and light according to need. Such “presence” or “occupancy” detectors required another set of specifications, and PIR detectors were adapted accordingly. Presence detectors are generally ceiling mounted and require a zone pattern to catch any sitting person in the room. Sensitivity is very high compared to alarm systems and has often an adaptive mode to increase it to the utmost possible. While a false alarm is annoying for an alarm system, it is of less importance for the presence detector. However, the presence detector should never miss someone and leave him in the dark.

Although the presence or occupancy detector for energy control is still in the beginning, some people consider this field as the largest potential for PIR applications, especially when such detectors are linked to building automation communication buses and perform an integral function for energy, security and comfort.

PIR seems to be the most effective “people detector”. PIR detectors are small, simple, cheap, very low power and do not emit anything. Unlike radar, ultrasonic or active (reflective beam type) infrared, PIR senses the body temperature as an additional criterion beyond the size and motion of the person. PIR detectors are especially useful in combination with a radio transmitter for wireless alarm systems, because the “passive” nature of PIR allows continuous operation for 10 years from a lithium battery. Hundreds of other applications have already been considered, from computers to coffee machines and toys.

7.5.3 General Discussion of Thermal IR Remote Sensing

Thermal infrared energy is emitted from all objects that have a temperature greater than absolute zero.Any object with a kinetic temperature greater than absolute zero (0K or −273℃) emits radiation whose intensity and spectral composition are a function of the material type and the temperature of the object under consideration. It is useful to describe thermal infrared radiation in terms of a hypothetical perfect target generally referred to as a blackbody. At any wavelength, a blackbody totally absorbs and re-emits all energy incident upon it. A blackbody behaves in a predictable manner as defined by the Stephan-Boltzmann law

$$M_{bb} = \sigma T^4 \tag{7.12}$$

where M_{bb} is radiant emittance in W/(cm^2 • μm); σ=5.6687×10^{-8}W/(m^2 • K^4) denotes Stephan-Boltzmann constant; T is absolute temperature.

Thus, over the entire range of the electromagnetic spectrum, the total energy emitted from an object is inversely related to the fourth power of the absolute temperature. The remote measurement of radiant emittance M_{bb} from a surface can, therefore, be used to infer temperature of the surface. Fundamentally, it is indirect approach to temperature measurement that is used in thermal remote sensing; i.e., radiant emittance M_{bb} is measured over a discrete wavelength range and used to find the radiant temperature of the radiating surface.

The spectral radiant emittance of a blackbody is defined by Planck's equation

$$L_{bb}(\lambda,T) = C_1 / \{\pi\lambda^5[\exp(C_2 / \lambda T) - 1]\} \tag{7.13}$$

where L_{bb} is blackbody radiance in W/(cm^2•μm); λ is wavelength in μm; T is temperature in K; C_1=3.7415×10^8 (W • μm^4/m^2); C_2=1.4338×10^4 (μm • K).

As seen by Planck's equation, reflectance of solar irradiation is of no significance in thermal IR remote sensing because at an estimated temperature of approximately 5900K, the emitted radiation from the sun (measured at the top of the Earth's atmosphere) is less than 1% of its radiation within the 8μm~12μm thermal window, which is where the vast majority of thermal remote sensing of earth's surface is conducted. The ambient temperature of the earth (i.e., the 'normal' temperature of natural surface materials such as soil, water, and vegetation) is about 300 K (27℃), and peak wavelength occurs around a wavelength of 9.7 μm. Because this radiation correlates with terrestrial heat, it is generally referred to as thermal infrared energy.

In the following, there are two examples about the application of the thermal infrared remote sensing: the first is the application of satellite thermal infrared remote sensing in monitoring magmatic activity of Changbaishan Tianchi volcano; the second is the application of remote sensing satellite data in the Study of Urban Population-Environment Interactions.

Magmatic activity is one kind of manifestation of both crustal movement and the interior earth heat energy release. Before eruption, the magma with high temperature ascends to the shallow crust along faults, cracks and volcanic channels during intensive volcanic activity stage or former volcanic eruption stage. As the result, the shallow crust, then the country rock, groundwater and the soil layers are heated up. Consequently, this process leads to the thermal anomaly which is strong enough to be detected by some methods. In addition, gases released from springs and cracks, such as CO_2, CH_4, bring up greenhouse effect, which causes local warming phenomena. In this way, sudden thermal anomaly and greenhouse effect will cause the volcanic area warming up clearly. Moreover, thermal anomaly is the main feature of intensive magmatic activity, and electromagnetic wave with a certain wavelength and certain distribution radiation energy caused by such anomaly can be received by thermal infrared sensors. Therefore, it is reasonable to believe that the thermal anomaly due to volcanic activity may be detected from the thermal infrared images.

Population modeling was one of the early applications of remote sensing (de Sherbinin et al. 2002). Although the number of people living in an areas can not be seen directly on the remotely sensed data, it can used as an indirect tool for population estimation by using different methods. The number of dwelling units can be multiplied by the average size of the household in obtaining the estimation of population from remote sensing data (aerial photographs and high resolution satellite data) using three different methods: (1) dwelling unit technique, (2) built-up areas technique and (3) housing density technique (Taragi et al, 1999). Another method, the land use area density method (multiplying urban built up areas by average population densities) was used in estimating the population of Bhimawaram town in India using IRS LISS-I data of 1988 (Raghavaswami, 1994). Remote sensing data also assists in planning censuses by identifying areas of new development and provides regular updates of new housing stock for planners.

7.6 Additive Manufacturing

The basic principle of Additive Manufacturing (AM) is that a model, initially generated using a three-dimensional Computer Aided Design (3D CAD) system, can be fabricated directly without the need for process planning. Although this is not in reality as simple as it first sounds, AM technology certainly significantly simplifies the process of producing complex 3D objects directly from CAD data. Other manufacturing processes require a careful and detailed analysis of the part geometry to determine things like the order in which different features can be fabricated, what tools and processes must be used, and what additional fixtures may be required to complete the part. In contrast, AM needs only some basic dimensional details and a small amount of understanding as to how the AM machine works and the materials that are used.

The key to how AM works is that parts are made by adding material in layers, each layer is a thin cross-section of the part derived from the original CAD data. Obviously in the physical world, each layer must have a finite thickness to it and so the resulting part will be an approximation of the original data, as illustrated by Fig. 7.23. The thinner each layer is, the closer the final part will be to the original. All commercialized AM machines to date use a layer-based approach, and the major ways that they differ are in the materials that can be used, how the layers are created, and how the layers are bonded to each other. Such differences will determine factors like the accuracy of the final part plus its material properties and mechanical properties. They will also determine factors like how quickly the part can be made, how much postprocessing is required, the size of the AM machine used, and the overall cost of the machine and process.

Fig.7.23 CAD image of a teacup with further images showing the effects of building using different layer thicknesses

7.6.1 The Generic Process of Additive Manufacturing

AM involves a number of steps that move from the virtual CAD description to the physical resultant part. Different products will involve AM in different ways and to different degrees. Small, relatively simple products may only make use of AM for visualization models, while larger, more complex products with greater engineering content may involve AM during numerous stages and iterations throughout the development process. Furthermore, early stages of the product development process may only require rough parts, with AM being used because of the speed at which they can be fabricated. At later stages of the process, parts may require careful cleaning and postprocessing (including sanding, surface preparation and painting) before they are used, with AM being useful here because of the complexity of form that can be created without having to consider tooling. Later on, we will investigate thoroughly the different stages of the AM process, but to summarize, most AM processes involve, to some degree at least, the following eight steps (as illustrated in Fig. 7.24).

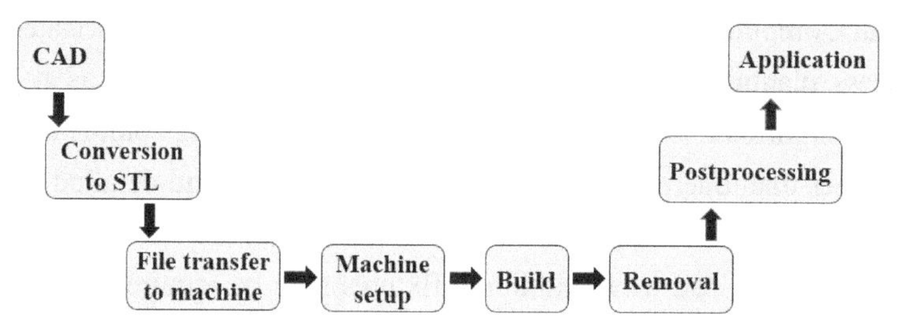

Fig.7.24 Generic process of CAD to part, showing all 8 stages

- **Step 1: CAD**

All AM parts must start from a software model that fully describes the external geometry. This can involve the use of almost any professional CAD solid modeling software, but the output must be a 3D solid or surface representation. Reverse engineering equipment (e.g., laser scanning) can also be used to create this representation.

- **Step 2: Conversion to Stereolithography (STL)** File Format

Nearly every AM machine accepts the STL file format, which has become a standard, and nearly every CAD system can output such a file format. This file describes the external closed surfaces of the original CAD model and forms the basis for calculation of the slices.

- **Step 3: Transfer to AM Machine and STL File Manipulation**

The STL file describing the part must be transferred to the AM machine. Here, there may be some general manipulation of the file so that it is the correct size, position, and orientation for building.

- **Step 4: Machine Setup**

The AM machine must be properly set up prior to the build process. Such settings would relate to the build parameters like the material constraints, energy source, layer thickness, timings, etc.

- **Step 5: Building**

Building the part is mainly an automated process and the machine can largely carry on without supervision. Only superficial monitoring of the machine needs to take place at this time to ensure no errors have taken place like running out of material, power or software glitches, etc.

- **Step 6: Removal**

Once the AM machine has completed the build, the parts must be removed. This may require interaction with the machine, which may have safety interlocks to ensure for example that the operating temperatures are sufficiently low or that there are no actively moving parts.

- **Step 7: Postprocessing**

Once removed from the machine, parts may require an amount of additional cleaning up before they are ready for use. Parts may be weak at this stage or they may have supporting features that must be removed. This therefore often requires time and careful, experienced manual manipulation.

- **Step 8: Application**

Parts may now be ready to be used. However, they may also require additional treatment before they are acceptable for use. For example, they may require priming and painting to give an acceptable surface texture and finish. Treatments may be laborious and lengthy if the finishing requirements are very demanding. They may also be required to be assembled together with other mechanical or electronic components to form a final model or product.

While the numerous stages in the AM process have now been discussed, it is important to realize that many AM machines require careful maintenance. Many AM machines use fragile laser or printer technology that must be carefully monitored and that should preferably not be used in a dirty or noisy environment. While machines are designed to operate unattended, it is important to include regular checks in the maintenance schedule, and that different technologies require different levels of maintenance. It is also important to note that AM processes fall outside of most materials and process standards. However, many machine vendors recommend and provide test patterns that can be used periodically to confirm that the machines are operating within acceptable limits.

In addition to the machinery, materials may also require careful handling. The raw materials used in some AM processes have limited shelf-life and must also be kept in conditions that prevent them from unwanted chemical reactions. Exposure to moisture, excess light, and other contaminants should be avoided. Most processes use materials that can be reused for more than one build. However, it may be that reuse could degrade the properties if performed many times over, and therefore a procedure for maintaining consistent material quality through recycling should also be observed.

7.6.2 Associated Technologies

- **Computers**

Like many other technologies, AM technologies came about as a result of the invention of the computer. AM takes full advantage of many important features of computer technology, both directly (in the AM machines themselves) and indirectly (within the supporting technology). Without computers there would be no capability to display 3D graphic images. Without 3D graphics, there would be no Computer-Aided Design. Without this ability to represent objects digitally in 3D, we would have a limited desire to use machines to fabricate anything but the simplest shapes. It is safe to say, therefore, that without the computers we have today, we would not have seen the development of AM technology.

- **Computer-Aided Design Technology**

Additive Manufacturing technology primarily makes use of the output from mechanical engineering, 3D Solid Modeling CAD software. It is important to understand that this is only a branch of a much larger set of CAD systems and, therefore, not all CAD systems will produce output suitable for layer-based AM technology. Currently, AM technology focuses on reproducing geometric form, and so the better CAD systems to use are those that produce such forms in the most precise and effective way. The demands on CAD technology in the future are set to change with respect to AM. As we move toward more and more functionality in the parts produced by AM, we must understand that the CAD system must include rules associated with AM. To date, the focus has been on the external geometry. In the future, we may need to know rules associated with how the AM systems function so that the output can be optimized.

- **Laser technology**

Many of the earliest AM systems were based on laser technology. The reasons are that lasers provide a high intensity and highly collimated beam of energy that can be moved very quickly in a controlled manner with the use of directional mirrors. Since AM requires the material in each layer to be solidified or joined in a selective manner, lasers are ideal candidates for use, provided the laser energy is compatible with the material transformation mechanisms. There are two kinds of laser processing used in AM: curing and heating. With photopolymer resins the requirement is for laser energy of a specific frequency that will cause the liquid resin to solidify, or "cure." Usually this laser is in the ultraviolet range but other frequencies can also be used. For heating, the requirement is for the laser to carry sufficient thermal energy to cut through a layer of solid material, to cause powder to melt, or to cause sheets of material to fuse. For powder processes, for example, the key is to melt the material in a controlled fashion without creating too great a build-up of heat, so that when the laser energy is removed, the molten material rapidly solidifies again. For cutting, the intention is to separate a region of material from another in the form of laser cutting. Earlier AM machines used tube lasers to provide the required energy but many manufacturers have more recently switched to solid-state technology, which provides greater efficiency, lifetime, and reliability.

- **Printing Technologies**

Ink-jet or droplet printing technology has rapidly developed in recent years. Improvements in resolution and reduction in costs has meant that high-resolution printing, often with multiple colors, is available as part of our everyday lives. Such improvement in resolution has also been supported by improvement in material handling capacity and reliability. Initially, colored inks were low viscosity and fed into the print heads at ambient temperatures. Now it is possible to generate much higher pressures within the droplet formation chamber so that materials with much higher viscosity and even molten materials can be printed. This means that droplet deposition can now be used to print photocurable and molten resins as well as binders for powder systems. Since print heads are relatively compact devices with all the droplet control technology highly integrated into these heads (like the one shown in Fig. 7.25), it is possible to produce low-cost, high-resolution, high-throughput AM technology. In the same way that other AM technologies have applied the mass-produced laser technology, other technologies have piggy-backed upon the larger printing industry.

- **Programmable Logic Controllers**

The input CAD models for AM are large data files generated using standard computer technology. Once they are on the AM machine, however, these files are reduced to a series of process stages that require sensor input and signaling of actuators. This is process and machine control that often is best carried out using microcontroller systems rather than microprocessor systems. Industrial microcontroller systems form the basis of Programmable Logic Controllers (PLCs), which are used to reliably control industrial processes. Designing and building industrial machinery, like AM machines, is much easier using building blocks based around modern PLCs for coordinating and controlling the various steps in the machine process.

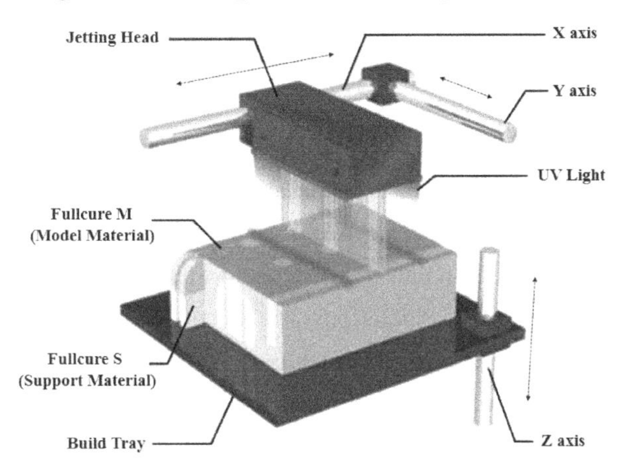

Fig.7.25 Printer technology used on an AM machine (photo courtesy of Objet)

- **Materials**

Earlier AM technologies were built around materials that were already available and that had been developed to suit other processes. However, the AM processes are somewhat unique and these original materials were far from ideal for these new applications. For example, the early photocurable resins resulted in models that were brittle and that warped easily. Powders used in laser melting processes degraded quickly within the machine and many of the materials used resulted in parts that were quite weak. As we came to understand the technology better, materials were developed specifically to suit AM processes. Materials have been tuned to suit more closely the operating parameters of the different processes and to provide better output parts. As a result, parts are now much more accurate, stronger, and longer lasting and it is even possible to process metals with some AM technologies. In turn, these new materials have resulted in the processes being tuned to produce higher temperature materials, smaller feature sizes, and faster throughput.

- **Computer Numerically Controlled Machining**

One of the reasons AM technology was originally developed was because Computer Numerically Controlled (CNC) technology was not able to produce satisfactory output within the required time frames. CNC machining was slow, cumbersome, and difficult to operate. AM technology on the other hand was quite easy to set up with quick results, but had poor accuracy and limited material capability. As improvements in AM technologies came about, vendors of CNC machining technology realized that there was now growing competition. CNC machining has dramatically improved, just as AM technologies have matured. It could be argued that high-speed CNC would have developed anyway, but some have argued that the perceived threat from AM technology caused CNC machining vendors to rethink how their machines were made. For geometries that can be machined using a single set-up orientation, CNC machining is often the fastest, most cost-effective method. For parts with complex geometries or parts which require a large proportion of the overall material volume to be machined away as scrap, AM can be used to more quickly and economically produce the part than when using CNC.

7.6.3 Business Opportunities and Future Directions

Because of recent developments in additive manufacturing, there is no fundamental reason for products to be brought to markets through centralized development, production, and distribution. Instead, products can be brought to markets through product conceptualization, product creation, and product propagation being carried out by individuals and communities in any geographical region.

Many companies already use the Internet to collect product ideas from ordinary people from diverse locations. However, these companies are feeding these ideas into the centralized physical locations of their existing business operations for detailed design and creation. Distributed conceptualization, creation, and propagation can supersede concentrated development, production, and distribution by combining AM with novel human/digital interfaces which, for instance, enable nonexperts to create and modify shapes. Additionally, body/place/part scanning can be used to collect data about physical features for input into digitally-enabled design software and onward to AM.

AM makes it possible for digital designs to be transformed into physical products at that same location or any other location in the world (i.e., "design anywhere, build anywhere"). Moreover, the web tools associated with Web 2.0 are perfect for the propagation of product ideas and component designs that can be created through AM. The combination of Web 2.0 with AM can lead to new models of entrepreneurship.

Distributed conceptualization and propagation of digital content is known as digital entrepreneurship. However, the exploitation of AM to enable distributed creation of physical products goes beyond just digital entrepreneurship. Accordingly, the term, digipreneurship was coined to distinguish distributed conceptualization, propagation and creation of physical products from distributed conceptualization and propagation of just digital content. Thus digipreneurship is focused on transforming digital data into physical products using an entrepreneurship business model.

Web 2.0 + AM has the potential to generate distributed, sustainable employment that is not vulnerable to off-shoring. This form of employment is not vulnerable to off-shoring because it is based on distributed networks in which resource costs are not a major proportion of total costs. Employment that is generated is environmentally friendly because, for example, it involves much lower energy consumption than the established concentration of product development, production, and distribution, which often involves shipping of products worldwide from centralized locations. As discussed throughout this section, developments in AM offer possibilities for new types of products. Thus, there are many potential markets for the outputs of digipreneurship.

7.7 Terahertz Techniques and Applications

The terahertz (THz) regime of the electromagnetic spectrum is broadly recognized by the frequency range of 100 GHz to 10 THz (where 1 THz corresponds to a frequency of 10^{12} Hz, a wavelength of 300μm, and photon energy of 4.1 meV). This region, alternatively called the far-IR, lies between the infrared light and the microwave frequencies, as shown in Fig. 7.26.

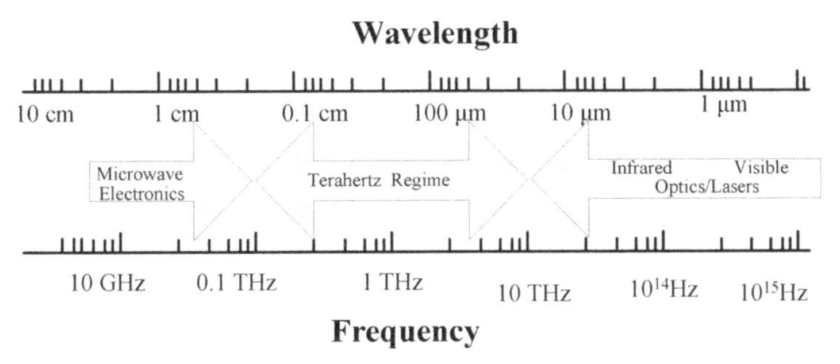

Fig.7.26 The THz gap—THz regime of the EM spectrum extends from 100 GHz to 10 THz, which lies below visible and infrared (IR) wavelengths and above microwave wavelengths.

The technical difficulties involved in making efficient and compact THz sources and detectors, and the lack of suitable technologies led to the THz band being called the "THz gap". This technological gap has been rapidly diminishing for the last decades. Optical technologies have made tremendous advances from the high frequency side, while microwave technologies encroach up from the low frequency side. This part gives a brief perspective on the basic properties of THz radiation and the progress of THz science and technology.

7.7.1 Terahertz Sources

Terahertz sources can be divided into two major categories: electronic and photonic sources. Terahertz electronic sources that are widely used include electron beam and solid-state sources, and frequency multipliers. The common terahertz photonic sources cover terahertz semiconductor and gas lasers and terahertz optoelectronic sources. The popular THz generation mechanisms are as follows.

- **Difference frequency generation**

One way of generating THz radiation is to exploit a nonlinear medium in which nonlinear frequency conversion is taken place for the incident electromagnetic waves, as shown in Fig. 7.27. Optical rectification and difference frequency generation (DFG) are second order nonlinear optical processes in which a THz photon at frequency ω_T is created by interaction of two optical photons at frequencies ω_1 and ω_2 with a nonlinear crystal, i.e.,$\omega_T=\omega_1-\omega_2$. Femtosecond laser pulses with a broad spectrum (bandwidth about 10 THz) generate broadband THz pulses by optical rectification, whose pulses shape is similar to the optical pulse envelope. Two CW optical beams produce CW THz radiation by DFG. Solid-state THz sources based on microwave technology convert incoming microwaves into their harmonic waves utilizing diodes with strongly nonlinear I-V characteristics.

- **Photomixing generation**

According to the principle of electrodynamics, the accelerating charges and time-varying currents will radiate electromagnetic waves. Figure 7.28 illustrates the THz radiation can be generated from a biased photoconductive (PC) antenna excited by laser beams. A PC antenna consists of two metal electrodes deposited on a semiconductor substrate. An optical beam illuminating the gap between the electrodes, generates photocarriers, and a static bias field accelerates the free carriers. This photocurrent varies with time, whose amplitude corresponds to the intensity of the incident laser beam. Consequently, femtosecond laser pulses produce broadband THz pulses. Mixing two laser beams with different frequencies forms an optical beat, which generates CW THz radiation at the beat frequency, which is called photomixing.

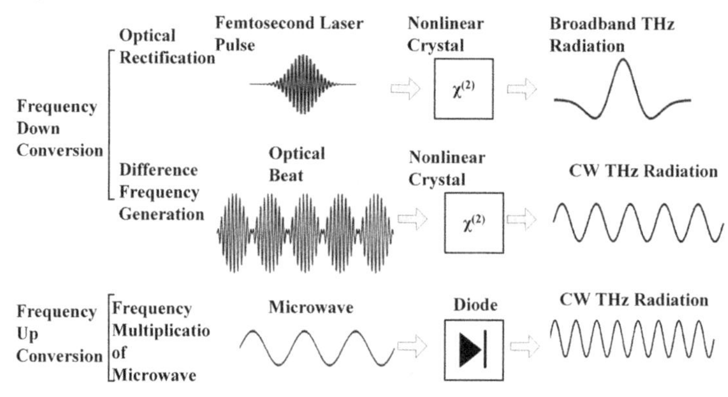

Fig.7.27 Terahertz generation in nonlinear media

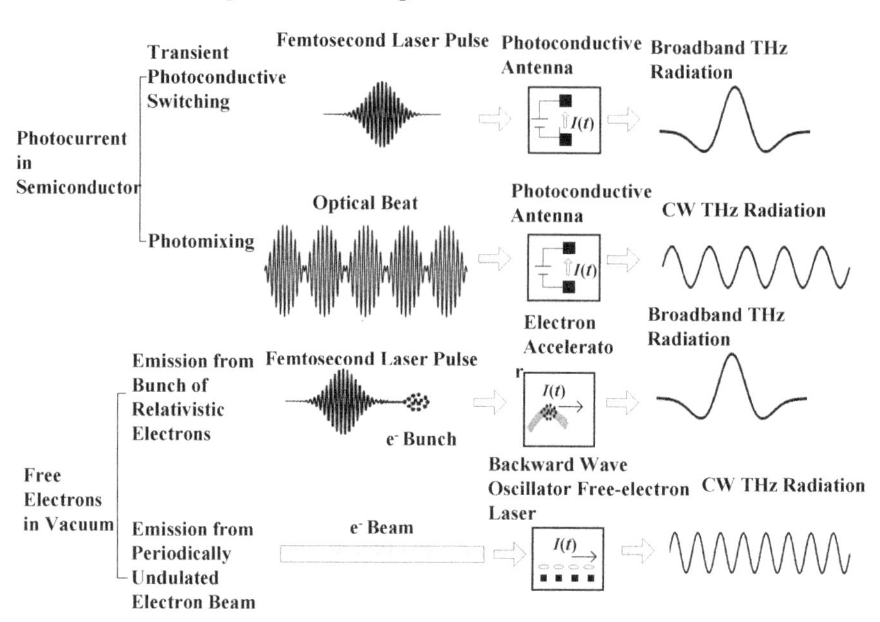

Fig.7.28 THz radiation from accelerating electrons

Electron accelerators produce extremely high-intensity THz radiation using relativistic electrons. A femtosecond laser pulse triggers an electron source to generate an ultrashort pulse of electrons. After being accelerated to a relativistic speed, the electrons are flushed into a metal target, or are forced into circular motion by a magnetic field. Coherent THz radiation is generated by this transient electron acceleration.

- **Photonic sources-THz gas lasers**

Several types of lasers have been developed for the THz region of the electromagnetic spectrum. They are characterized by underlying quantum mechanical transitions between different energy levels. For instance, traditional molecular gas lasers are based on the transitions between rotational modes of molecules, and solid state lasers such as quantum cascade lasers and p-type germanium lasers rely on intraband transitions in semiconductors.

The key process governing radiation from these different types of lasers is stimulated emission. Figure 7.29 illustrates the basic concepts of how a typical laser works. The elemental parts of a laser system include a gain medium, a laser cavity, and a pump. The gain medium is a material system in which stimulated emission takes place. Population inversion, which means that more atoms or molecules in the gain medium are in the excited states than the lower energy states, is a prerequisite of stimulated emission. Therefore, a pump source is necessary to maintain the system in the high energy state. Light is confined within the laser cavity by high reflectors. One of the reflectors is partially transmissive so that its transmitted radiation can be used as the laser output. The light confinement leads to amplification of the radiation intensity within the cavity, which encourages stimulated emission.

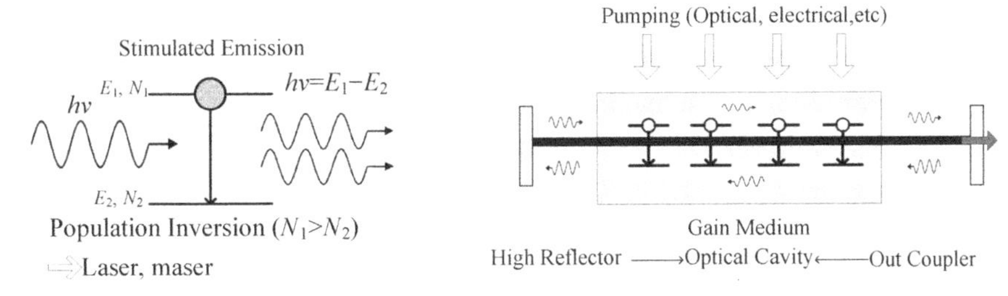

Fig.7.29 Schematics of a typical laser operation

As stated previously, a population inversion is imperative for laser operation. It, however, cannot be attained in a two-level system by any pumping schemes because, for an external perturbation, the rate of the transition from the lower to the higher level is exactly the same as the one from the higher to the lower level. Additionally, spontaneous emission is an extra pathway for the higher-to-lower transition. Therefore, how hard the system is pumped, an equal population of the two levels is as good as it gets.

The basic design of THz gas lasers is similar to that of the typical laser system. An extra component of importance is an intracavity waveguide used to confine the laser modes in the transverse direction. The gain media of THz gas lasers are molecular gases such as CH_3F, CH_3OH, NH_3 and CH_2F_2. The THz radiation originates from the rotational transitions of the molecules. The molecules have permanent dipole moments, hence their rotational transitions are directly coupled to electromagnetic radiation via dipole interactions.

Figure 7.30 illustrates the lasing scheme of a typical THz gas laser. Optical pumping with a CO_2 laser excites some of the molecules from the lowest to the first excited vibrational mode. For symmetric-top molecules, the vibrational-rotational transitions obey the selection rules $\Delta v = 1$, $\Delta J = 0$ or ± 1, and $\Delta K = 0$. The optically induced population inversions between $(J+1)$ and J-levels for $v = 0$ and between J and $(J-1)$ levels for $v = 1$ give rise to emissions at THz frequencies. The cascade transition from $(J-1)$ to $(J-2)$ level for $v = 1$ also contributes to the THz radiation.

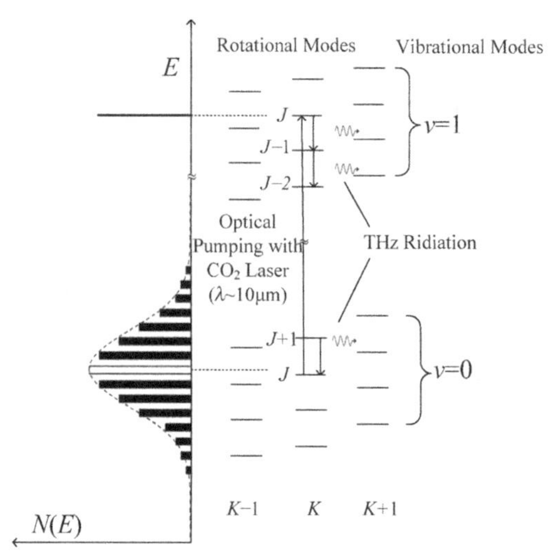

Fig.7.30 Energy level diagram of optical excitation ($v = 0 \rightarrow 1$) and THz radiation ($J+1 \rightarrow J$ for $v = 0$, $J \rightarrow J-1$ and $J-1 \rightarrow J-2$ for $v = 1$) in an optically-pumped THz gas laser.

Many chemical species have been examined for lasing in the THz region, and several hundred THz laser emission lines have been observed. Table 7.3 lists some of the stronger laser lines in the THz region.

Tab.7.3 Laser lines of optically pumped THz gas lasers

Frequency(THz)	Molecule	Output Power(mW)
8.0	CH_3OH	~10
7.1	CH_3OH	~10
4.68	CH_3OH	>20
4.25	CH_3OH	~100
3.68	NH_3	~100
2.52	CH_3OH	>100
2.46	CH_2F_2	~10
1.96	$^{15}NH_3$	~200
1.81	CH_2F_2	<100
1.27	CH_2F_2	~10
0.86	CH_3Cl	~10
0.59	CH_3I	~10
0.525	CH_3OH	~40
0.245	CH_3OH	~10

7.7.2 Terahertz Detectors

THz detection schemes are generally classified as either coherent or incoherent techniques. The essential difference is that coherent detection measures for both the amplitude and phase of the field, whereas incoherent detection measures only for the intensity. Coherent detection techniques are closely associated with generation techniques in that they share underlying mechanisms and key components. In particular, optical techniques utilize the same light source for both generation and detection.

- **Coherent Detectors**

Figure 7.31 illustrates the commonly used coherent detection schemes. Free space electro-optic (EO) sampling measures the actual electric field of broadband THz pulses in the time domain by utilizing the Pockels effect, which is closely related to optical rectification. A THz field induces birefringence in a nonlinear optical crystal which is proportional to the field amplitude. The THz pulse shape is determined by a weak optical probe measuring the field induced birefringence as a function of the relative time delay between the THz and optical pulses.

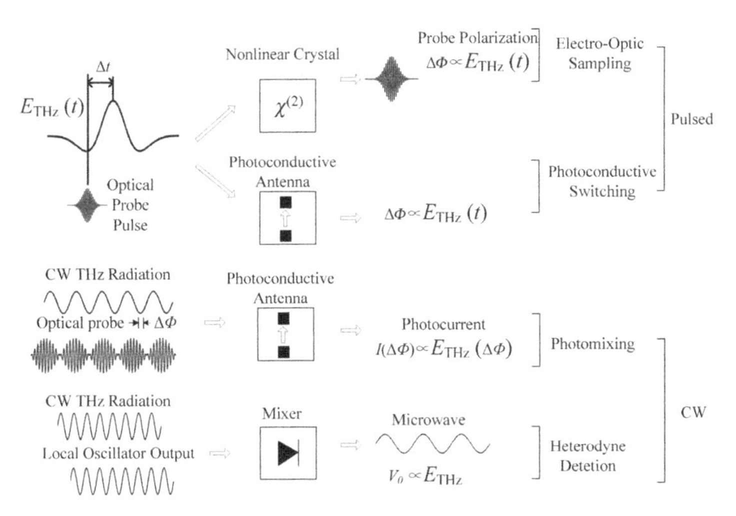

Fig.7.31 Coherent detection of THz radiation

The PC antenna also can be used to measures broadband THz pulses in the time domain. In the absence of a bias field, a THz field induces a current in the photoconductive gap when an optical probe pulse injects photocarriers. The induced photocurrent is proportional to the THz field amplitude. The THz pulse shape is mapped out in the time domain by measuring the variation of the photocurrent with the time delay between the THz pulse and the optical probe.

Photomixing measures CW THz radiation by exploiting photoconductive switching. In this case, the photocurrent shows sinusoidal dependence on the relative phase between the optical beat and the THz radiation。

Heterodyne detection utilizes a nonlinear device called a "mixer". Schottky diodes are commonly used as mixers. The key process in a mixer is frequency downconversion, which is carried out by mixing a THz signal ω_s with reference radiation at a fixed frequency ω_{LO}. The mixer produces an output signal at the difference frequency called the "intermediate frequency", $\omega_D=|\omega_s - \omega_{LO}|$. The amplitude of the output signal is proportional to the THz amplitude. Unlike the optical techniques, heterodyne detection is usually used to detect incoherent radiation.

- **Incoherent Detectors**

Commonly used incoherent detectors are thermal sensors, such as bolometers, Golay cells, and pyroelectric devices, which is commonly used for observation of CW THz radiation. A radiation absorber attached to a heat sink is the basic element of all thermal detectors. Radiation energy absorbed by the absorber is converted into heat, and a thermometer measures the temperature increase induced by the heat. The absorber has a low heat capacity so that the heat brings about acute temperature changes. Each type of thermal detector is distinguished by the specific scheme that uses to measure the temperature difference between the absorber and the heat sink. The absorbed radiation energy is determined by calibrating the measurement output.

Bolometers are equipped with an electrical resistance thermometer made of a heavily doped semiconductor such as Si or Ge. In general, bolometers operate at cryogenic temperature. Pyroelectric detectors employ a pyroelectric material in which temperature change gives rise to spontaneous electric polarization. In a Golay cell, heat is transferred to a small volume of gas in a sealed chamber behind the absorber. An optical reflectivity measurement detects the membrane deformation induced by the pressure increase. These thermal detectors respond to radiation over a very broad spectral range. Because a radiation absorber must reach to thermal equilibrium for a temperature measurement, detection response is relatively slow compared with typical light detectors.

7.7.3 Terahertz Metamaterials

Typically, isotropic materials can be characterized by their effective dielectric (permittivity, ε) and magnetic (permeability, μ) properties. A ubiquitous material in nature is free space or air, with a permittivity of ε_0 and permeability of μ_0. The relative permittivity and permeability of a material are defined as $\varepsilon_r = \varepsilon/\varepsilon_0$ and $\mu_r = \mu/\mu_0$, respectively, from which the refractive index of that material can be written as $n=\sqrt{\varepsilon_r\mu_r}$. Figure 7.32 illustrates the classification of isotropic materials in the ε-μ space. In Fig. 7.32, the first quadrant ($\varepsilon > 0$ and $\mu > 0$) represents right handed materials (RHMs), which support the right-handed (forward) propagating waves in most dielectric or optical materials. From Maxwell's equation, the electric field $\boldsymbol{E}$, the magnetic field $\boldsymbol{H}$, and the wave vector $\boldsymbol{k}$ form a right-handed triplet.The second quadrant ($\varepsilon < 0$ and $\mu > 0$) represents the electric plasma where the incident EM wave decays and supports evanescent waves. Many metals in the ultra-violet and visible frequency range fall in this quadrant. The third quadrant ($\varepsilon < 0$ and $\mu < 0$) represents the left-handed materials (LHMs) which support many exotic electromagnetic properties, such as backward propagating waves, as predicted by Veselago in 1968. There are no known natural materials that exhibit properties of this quadrant. In LHM, the $\boldsymbol{E}$ - $\boldsymbol{H}$ -$\boldsymbol{k}$ triplet is given by the left-hand rule. The fourth quadrant ($\varepsilon > 0$ and $\mu < 0$) denotes the magnetic plasma, which supports evanescent waves and very few natural ferromagnetic materials at sub-GHz frequencies fall in this category.

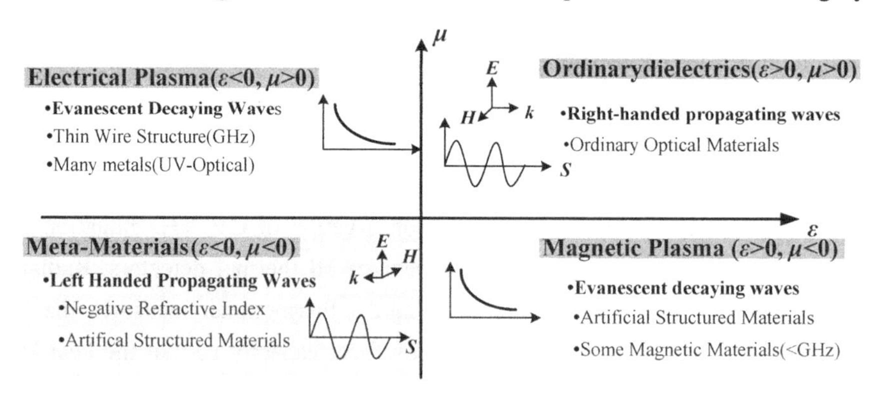

Fig.7.32 Classification of materials based on dielectric (ε) and magnetic (μ) properties. First quadrant represents materials with $\varepsilon > 0$ and $\mu > 0$, the second represents $\varepsilon < 0$ and $\mu > 0$, the third denotes $\varepsilon < 0$ and $\mu < 0$ and the fourth represents $\varepsilon > 0$ and $\mu < 0$

Metamaterials are macroscopic composite of periodic or non-periodic subwavelength structures, whose EM property is primarily a function of the cellular structure allowing great flexibility in creating new effective materials, unavailable in nature. The sub-wavelength inclusions are like the atomic particle in conventional material and the EM response of the metamaterial is characterized by the effective permittivity ε and permeability μ, determined in the homogenization of the periodic inclusions by averaging the local fields. In the early days, the term metamaterial referred to negative-refractive index material (NIM), the third quadrant in Fig. 7.29, phenomena first postulated by Veselago in 1968. He theoretically predicted materials with simultaneous negative values of the ε and μ, hence giving an index of refraction $n=\sqrt{\varepsilon_r\mu_r}$ that is less than zero, a material not naturally found even today. Along with negative-refractive index, Veselago also predicted other exotic properties such as reverse Cerenkov radiation, reverse Doppler shift, and opposite phase and group velocity, among others. In its long history, metamaterials, Left-Handed materials (LHMs), NIMs, double-negative materials (DNGs), and backward-wave materials have been regarded as the same terms. But today, the term metamaterials has a much broader scope than LHMs. They constitute any artificial EM materials constructed from sub-wavelength periodic or non-periodic inclusions resulting in a user defined effective ε, μ or both. In terms of the classification in Fig.7.32, today metamaterials denote any artificial effective medium that represents quadrant two, three and four. Moreover, many materials do not respond to THz radiation naturally, it is necessary then to build the electromagnetic devices which enable the construction of useful applied technologies operating within this range. These are devices such as directed light sources, lenses, switches, modulators and sensors. Since the first realization of terahertz metamaterials, nearly all metamaterial schemes, exhibiting either electric, magnetic, or both responses, have been scaled to and studied at the terahertz regime.

7.7.4 Terahertz Technologies for Applications

The THz region is crowded by innumerable spectral features associated with fundamental physical processes such as rotational transitions of molecules, large-amplitude vibrational motions of organic compounds, lattice vibrations in solids, intraband transitions in semiconductors, and energy gaps in super-conductors. THz applications exploit these unique characteristics of material responses to THz radiation.

THz radiation can penetrate dielectric materials, including plastics, ceramics, crystals and colorants, similar to electric waves, allowing it to be applied to transmission or tomographic imaging. In addition, spectrum information can be acquired using THz waves, similar to infrared technologies. Using transparency or spectral information features, THz applications are expected to improve as research advances in numerous sectors, including manufacturing, pharmaceuticals, materials, semiconductors and security.

- **Terahertz applications in tomographic imaging and material spectroscopy**

Microscopic imaging can provide structural and functional information from preserved tissue specimens, but is used mainly in histopathology. Conversely, medical imaging methods, including magnetic resonance imaging (MRI) and computed tomography (CT), provide macroscopic images of living tissues, albeit with much lower resolution and specificity. THz-wave technology is expected to resolve these issues, due to its efficient resolution of medical imaging and tissue spectral information. Currently, imaging living tissue with high water content is difficult. However, adding THz-wave technology to medical imaging tools could increase the output power and signal-to-noise ratio (SNR) of the systems. Moreover, tomography using THz waves has been developed for non-destructive investigations, including in aerospace devices and art. Tomographic imaging has also been introduced in industrial applications, such as automobile part examinations and quantitative analyses of powdered chemicals. Tomography can acquire three-dimensional (3D) images within the THz frequency, similar to the photonic or X-ray regions of the electromagnetic spectrum.

- **Terahertz applications in the aerospace industry**

Glass fiber reinforced plastics (GFRP) and carbon fiber reinforced plastics (CFRP) are increasingly being used as structural components in aircraft, because of their high strength-to-weight ratios, improved aerodynamic performance and reduced corrosion compared with other structural materials. Composites can be weakened by various defects and stress during the lifecycle of an aircraft, and routine maintenance of composite materials requires rather complicated inspection and repair techniques. THz radiation has the unique ability to penetrate composites and identify defects such as voids, delaminations, mechanical damage or heat damage. THz offers a non-invasive, non-contact, non-ionizing method of assessing composite part condition and can overcome some of the shortcomings of other non-destructive techniques such as X-rays, ultrasound, video inspection, eddy currents and thermographic techniques. THz time domain spectroscopy (TDS) has been investigated as a possible method of quality control of polymeric compounds and their composites by imaging. THz, in both continuous-wave (CW) and pulsed modes, has also been studied as a means of evaluating damage to carbon fiber composites. A THz imaging system can provide a non-destructive, standoff imaging technique capable of detecting corrosion on metallic surfaces under obscurants, and defects in composite materials on aerospace structures

The outer skin of the aircraft consists of thin (~ millimeter) sheets of fiber composites that are bonded to honeycombed composites for structural support. During the manufacturing process, quality-assurance inspections could be performed on these thin sheets using a transmissive THz system. A transmissive system can accurately determine the optical path length of the sheet which provides useful information regarding the epoxy homogeneity and regions of delamination. For inspections after the aircraft has been manufactured, reflective THz geometry is required because there are often structural components beneath the layer of interest that obstruct the penetration of the transmitted THz wave.

- **Terahertz wireless communications**

The rapidly increasing demand for higher bandwidth and data rates in wireless communication systems over the past several decades is well known. According to Edholm's law, the demand for point-to-point bandwidth in wireless short-range communications has doubled every 18 months over the last 25 years. It can be predicted that data rates of around 5~10 Gb/s will be required in ten years. Some of the reported advantages of THz communication links compared to millimeter-wave links are inherently higher utilizable bandwidth relative to the higher carrier frequency, less susceptibility to scintillation effects than infrared wireless links, and the ability to use THz links for secure communications.

In conjunction with the high-bandwidth potential, THz links will exhibit an intrinsically short path length and line-of-sight communication. In the discussion of future THz communication systems, some researchers suggests that the commercial application of THz communication links would be a niche in which very high data rates are required over short distances on a multi-point-to-point/multi-point basis (i.e., the 'first' and 'last mile' problem). The 'last' and 'first' mile problem refers to establishing broadbanded, multi-user local wireless connections to high-speed networks

The intrinsic advantage of THz communication systems compared to microwave or millimeter-wave systems is that of higher bandwidth. However, what about the other competing frequency limit, namely infrared (IR) free-space communications? IR free-space communication links at 1.5 μm wavelength are the most common optical transmission vehicle for short reach (up to 10 km). Wireless IR systems are 30 years old, yet until recently the highest data rates reported were 155 Mb/s. A 2007 review of the field showsno improvement beyond the 155 Mb/s data rate reported in 2001. Only recently has a 10 Gb/s data rate been demonstrated in a simulated atmospheric environment. The key to increasing the IR wireless data rate to10 Gb/s was advanced modulation formats such as orthogonal frequency division multiplexing.

- **Terahertz bio-sensing techniques**

Because the characteristic energy of biological materials, generated by molecular motions such as rotation and vibration, lies in the THz frequency range, THz spectroscopy is the best tool for the direct sensing of biological molecules and the associated techniques.

THz waves have been utilized for the study of biological and medical materials, revealing many new phenomena and demonstrating the feasibility of their further use in diagnosing diseases such as cancer. Investigating biological characteristics and dynamics using THz techniques has both advantages and disadvantages. One of the most important advantages is the low energy of THz waves, corresponding to a few meV, which is well below the ionization energies of atoms and molecules. This means that we can study materials with THz waves without disturbing the system under study, unless enough power is applied to cause a significant increase in temperature. The other major advantage is that the characteristic energy of the hydrogen bond, which is the most dominant bond in biological molecules, lies in the THz frequency range and, therefore, THz waves can directly detect spectral features such as resonances and motions of molecules. However, a serious disadvantage is the high absorption of THz waves by the water in which biological molecules reside. This water absorption can mask the characteristic features of samples, even though biological dynamics naturally occur in water.

- **Terahertz applications in the pharmaceutical industry**

The THz region of the electromagnetic spectrum spans the frequency range between the mid-infrared (mid-IR) and the millimetre/microwave. The centre portion of the THz region (0.1~4 THz, 3.3~133 cm^{-1}) offers a number of very useful properties desired for pharmaceutical applications: (1) THz radiation gives rise to individual 'fingerprints' spectra for many crystalline materials including active pharmaceutical ingredients (API), making THz spectroscopy a useful tool for pharmaceutical material characterisation. (2) THz radiation can penetrate deep into pharmaceutical tablets as most common pharmaceutical amorphous materials are semitransparent to THz radiation, thus THz imaging can be used to map internal structures of pharmaceutical products. (3) The photon energy of THz radiation is millions of times smaller than that of X-rays, thus it is safe to use as it will be unlikely to cause any damage to the sample under study. The combination of these properties makes THz spectroscopy and imaging a potentially very powerful technique for characterising the physical structures and chemical compositions of pharmaceutical solid dosage forms.

Spectroscopy in near- and mid-IR region has been widely used by the pharmaceutical industry, however, spectroscopy measurement in THz region has historically been difficult, owing to the lack of any suitable source and detector. With the advances in ultrafast laser technology, the past ten years have seen a revolution in THz devices and systems. Of particular significance is the development of THz time-domain spectroscopy (THz-TDS) and THz time-domain imaging (THz-TDI) systems. One of the most distinct features of THz-TDS and THz-TDI techniques is the coherent generation and detection of short pulse broadband THz radiation by using an ultrafast femtosecond laser system. There are three main advantages to using pulsed THz radiation and the associated coherent-detection scheme. (1) This technology directly measures the transient electric field, not simply the intensity of the THz radiation. This yields THz spectra with far better signal-to-noise ratio and dynamic range compared with the Fourier Transform Infrared (FTIR) method. High-quality THz spectra are now routinely obtained in less than 20 ms without the need for a cryogen-cooled bolometer, making THz spectroscopy more easily and widely accessible. (2) Because of the time-gated coherent-detection technology used, the extraneous ambient noise (originating from the incoherent blackbody radiation from the sample and its surroundings) is minimised. (3) The use of pulsed radiation and the associated coherent-detection scheme preserves the time-gated phase information, upon which THz imaging has been developed, for characterising the internal structures of a sample quantitatively and non-destructively.

Words and Expressions

a stack of	一堆，许多
active shutter 3D	主动快门式 3D 技术
actuator	促动器
adhesion	黏附，黏着力
affinity	亲和力
agent	试剂药剂
albeit	虽然，即使
align	使成一直线，对准
all-encompassing	包罗万象的
ambient	周围的，氛围的
ambiet	反光参数
ampere	安培
anaglyphic 3D	色差式 3D 技术

analogous	相似的
by analogy with	从……类推
analyte	被分析物
anatomical	结构（上）的，解剖的
anisotropic	各向异性的
anneale	退火处理
anomaly	异常，反常
antenna	天线
antibody	抗体
antireflection	减反射（增透）
aqueous	含水的
artificial	人造的，人工的
ashing	砂磨，抛光
astable	不稳定的
autostereoscopic	自由立体
additive manufacturing	增材制造
automated process	自动化处理
ambient temperature	环境温度
avenue	大街
bacteria	细菌
benign	善良的，良性的
bioadhesives	生物胶带
biochip	生物芯片
biocompatibility	生物适应性
biodetection	生物探测
biofabrication	生物装配
bioimaging	生物成像
bioinformatics	生物信息学
biophotonics	生物光子学
biorecognition	生物识别
biosensor	生物传感器
bioterrorism	生化恐怖主义
biotronics	生物环境调节技术
birefringence	双折射
blackbody	黑体
bolometer	辐射热测定器
boundary	边界
bracelet	手镯，手环
brittle	易碎的，脆弱的
bulk	块，体积

bump	隆起，凸起
burglar	窃贼
cascade	串联，级联
CAT scan	计算机断层扫描技术
cellular	细胞的
census	人口调查，统计数字
ceramic	陶瓷
chamber	腔，室
clinic	诊所，临床
cluster	束，组，群，集，丛
collimate	校准，瞄准
colour division	色分复用
colorants	色素，着色剂
computer-assisted tomography	计算机辅助断层析术
compatible	兼容的
concentration	浓度
conceptualization	概念化
condensation	冷凝，凝缩，缩合，凝聚
condense	冷凝
confinement	限制
consciousness	意识，觉悟
contaminate	污染
contender	竞争者
convergence	收敛性
corrosion	腐蚀，衰败
cosmetic	整容的，表面的
creep	塑流，蠕动
crush	挤压
cryogenic	低温学的
cryopump	低温泵
cumbersome	笨重的，繁杂的
cunning	狡猾的，熟练的
cytometry	血细胞计数
maintenance	维护
contaminants	污染物
chemical reaction	化学反应
cumbersome	繁杂的，笨重的
deflect	偏转，偏移
debate	争论，辩论
deformation	变形

delamination	分层
deposition	沉积
developing	显影
diagnostic	诊断的，判断的
diffusion pump	扩散泵
digiproneurship	数码创业
diodes	二极管
disintegrate	分解
dome	圆顶盖，圆顶
dominant	主导的，突出的
dope	向…内掺添加物
dosage	剂量
downconversion	下转换
droplet	液滴
dwelling	居住，住所
eddy	涡流，漩涡
electrode	电极
electroluminescent	电致发光的
elicit	诱出，引出
emerging	新兴的
emission	发射，散发
encroach	蚕食，侵占
endogenous	内源性，内生性
endoscope	内窥镜
entrepreneurship	企业家精神
envelope	包络
enzyme	酶
etching	蚀刻术
epifluorescence excitation	反射荧光激发
epoxy	环氧的
equilibrium	均衡
etching	蚀刻
euphoria	鼓舞
euphotic	透光
evanescent	倏逝
evaporation	蒸发，消失
EVU lithography	极紫外线光刻技术
ex vivo	体外
excitedstate	激发态
exotic	外来的，独特性

exponentially	以指数方式
exposure	曝光
extinction coefficient	消光系数
external geometry	外部结构
fabricate	建造
farsightedness	远视
femtosecond	飞秒
ferromagnetic	铁磁的，铁磁体
filament	灯丝，细丝
filter wheel	滤光片转盘
fingerprint	指纹
fixture	夹具
glitches	小故障
fluorescence	发荧光，荧光
fluorochrome	荧光剂
fluorophore	荧光体
frames	帧
frontier	边界
full HD	全高清
gel	胶质体
genomics	基因组学
genotyping	基因型
germanium	锗
gram-negative rod-shaped bacterium	革兰式阴性杆状细菌
hamper	妨碍，阻止
harmonic	谐波
harness	利用
heterodyne	外差法，外差振荡器
histopathology	组织病理学
homogeneity	同质，同种
homogeneous	均匀的
honeycomb	使成蜂巢状
hypothetical	假想，假设
immunophenotyping	免疫表现型
immersion	侵入，浸没
imperative	必要的，不可避免的
implant	移植
inception	开始，开端
in vitro	【医】离体，在体外
indispensable	不可缺少的

infectious	传染的，有传染性的
ingredients	原料，成分
inorganic	无机的，无生物的
interlace	交织，组合，交错
interocular	两眼间的
intraband	带内
intracellular	细胞内的
intravenously	静脉注射
intrusion detectors	入侵探测器
invasive	扩散性
ion	离子
ionizing	电离的
irradiation	放射，照射
irregularities	不平整（度），不规则性
isolation	隔离
isotropic	各向同性的
iteration	重复，重述，反复，迭代
interlock	联动装置
ink-jet	喷墨式
kinetics	动力学
lenticular lens	双凸透镜
lattice	晶格
lifecycle	生活周期，生命循环
lifespan	寿命
ligand	配体
light-activated	光活化
likewise	同样地
lithography	光刻技术
low-defectivity	低缺陷
lung	肺
low viscosity	低黏度
macroscopic	肉眼可见的，宏观的
magmatic	岩浆的
magnetic resonance imaging（MRI）	磁共振成像
malignant	恶性的
manifestation	显示，表现
manipulation	操作，操控
membrane	膜，薄膜
MEMS-fabricated	微电子机械系统制造的
micelle	微包，微团

microarray	微阵列法
microbe	微生物
microcontroller	微控制器
microfabrication	精密加工
microorganism	微生物
micropatterning	缩微成像
microprocessor	微处理器
microscopy	显微镜
millennium	千年，千禧年
mingle	混合，混杂
miniaturize	使小型化
modalities	样式，形式，形态，感觉
modulator	调制器
monitoring	监视，监控
moisture	湿气，潮气，潮湿
molding	模压，压模，模塑
monochromatic	单频的，单色的
muddle	弄乱，混淆
multidimensional	多方面的，多维的
multidisciplinary	包括各种学科的
multiparameter	多参数
melt	融化
molten material	熔融物
moisture	水分，湿气
nanobioscience	纳米生物科学
nanoclinic	纳米诊所
nanocomposite	纳米化合物
nanodomain	纳米区
nanoimprint	纳米压印
nanomania	纳米狂热
nanomedicine	纳米药物
nanomer	纳米子
nanoparticle	纳米颗粒
nanophosphores	纳米荧光体
nanoshell	纳米外壳
nanosubmarine	纳米潜艇
nanotray	纳米支架
nasal	鼻腔的
non-invasive	非侵入性的
non-ionizing	非电离的

nonthermal	非热能的
occlude	封闭，阻塞
off-shoring	离岸，境外生产
oligonucleotide	低（聚）核苷酸
omit	省略，忽略
opaque	不透明
optimize	使最优化，使完善
organic	有机（体）的，有机物的
orthogonal	正交的，直角的
oxidize	氧化
pan-chromatic	全色的
panoramic	全景的
parallax barrier displays	视差栅栏显示
passband	通带
pellicle	薄皮，薄膜，胶片
perspective	观点，远景，透视图
penetrating	渗透，浸入
perception	感觉，知觉，获取
permanent	永久的，不变的
permeability	磁导率，导磁系数
perpendicular	垂直的，正交的
perturbation	扰动
pestis	鼠疫，黑死病
pharmaceutical	药物
phenomenon	现象
phenotyping	表现型
photocurrent	光电流
photo-dynamic	光动力学
photodynamic therapy	光动力疗法
photofabrication	光加工
photofrin	光敏素
photolithography	光刻
photomixing	光混频
photon	光子
photonic	光子的
photopolymerization	光致聚合
photoprocess	光学加工
photoresist	光刻胶，光敏抗蚀剂
photosensitizer	光敏剂
photosynthesis	光合作用

physical vapor deposition（PVD）	物理气相沉积法
piggy-backe	肩负，担任
pioneer	先锋，开拓者，先驱
pixels	像素
plasma	等离子体
plasmon	等离子体基元
Pockels effect	泡克耳斯效应
polarization	偏振
polymeric	聚合的
porous	可渗透的，多孔的
postulate	假设
powder	粉末
practitioners	从事者，实践者
prebake	预焙，焙烘
prerequisite	先决条件
probe	探针
proliferation	增殖，分芽繁殖，扩散
proportional	成比例的
prosthetic	修复学
proteomics	蛋白质组学
proximity	邻近，接近
pulse	脉冲
pyroelectric	热电的
punch	开洞，打洞
postprocessing	后处理
priming	底漆
photopolymer resins	感光性树脂
photocurable	光固胶
propagation	传播
quadrant	象限
quantum	量子论
radiography	放射线照相术
radioisotope	放射性同位素
reactive	活性的
rear	后部的，背后的
rectification	整流
reflectance	反射率
relief	救济
reinforce	加强，加固
remote sensing	遥感

resistance	阻力，电阻
resin	树脂，松香
reticule	标度线，刻线，分划板
retinal	视网膜
reuse	重复使用
ridge	背脊，隆起物
ripple	波纹（多用复数）
resin	用树脂涂
roadmap	线路与零件之布局图
rod	杆，棒
rough	未经加工的，粗糙的
rotate	旋转，（使）转动，（使）轮流
raw materials	原材料
saliva	唾液
scintillation	闪烁，发出火花
scrap	碎片，残余物
seal	密封
secretion	分泌物
shrinking	萎缩
shutter	快门
simultaneously	同时的
sinusoidal	正弦曲线的
situ	【医】原位
soluble	可溶的
solvent	溶剂
sophisticated	高级的
specimens	样本
spectroscopy	光谱学
spin coating	旋转涂胶
spontaneous	自发的，自然的
stain	染色，污点，污迹
stereo images	立体图像
stereogram	实体图
stereoscopic	有立体感的
stopband	禁带
stripper	去层器，去膜剂
sublime	升华，蒸升
submicron	亚微细米，亚微细粒
substrate	培养基
superficial	表面的，肤浅的

surgeries	外科手术
symmetric	对称的，匀称的
synthesize	合成
supersede	取代，代替
supervision	监督
software glitches	软件小故障
surface texture	表面纹理
template	模板，样规
terrestrial	陆地
theoreticians	理论家
therapy	疗法
thermographic	热成像的
thermometer	温度计
thin film	薄膜
thinning agent	稀释剂
throughput	生产量，容许量
tissue	组织
tomography	X 线断层摄影术
toxin	毒素
transducer	传感器
transmittance	透射率
tremendous	极大的
tri-methylated	三甲基
triplet	三个一组
tumor	肿瘤
turbomolecular pump	涡轮分子泵
ubiquitous	普遍存在的
ultrasonic	超声波
ultrasound imaging	超声波图像诊断
undercutting	低切馏出物
uppermost	最高的
urine	尿
utilize	利用
ultraviolet	紫外的
vacuum	真空
vague	含糊的，不清楚的
vibrational	振动的
virus	病毒
viscous	黏性的，黏的
visual cortex	视觉皮层

vitro 体外的，在体外
vivo 活的，活跃的
volcanic 火山的
vendor 供应商
viscosity 黏性
wafer 晶片
water repellent 防水处理
wax 蜡
weed out 除去
welding 粘接
wireless 无线的，无线电的
wrinkle 皱纹
warped 弯曲的
Y. pestis F1 鼠疫杆菌 F1
yeast 酵母

Grammar 专业英语翻译方法（七）：介词和介词短语的译法

介词的用法和译法都是习惯成自然，有些很难说出理由，只能强记。有些介词的用法和译法是很微妙的。例如，at，in 都有“在”的意思，但使用起来不同。

—I’ll be back at two o’clock.

我将在两点钟时回来。

—I’ll be back in two hours.

我将在两小时后回来。

林肯演说词：

—That Government of the people, by the people, for the people, should not perish from the earth.

民有民治民享的政府一定万古长青。

介词的翻译方法：

（1）转译：将英语介词译成汉语动词或形容词。

—He is for the suggestion, but I am against.

他支持这建议，但我反对。

—He went to the shop for a piece of paper.

他去商店买报纸。

—The president took the foreign guests about campus.

大学校长带着外宾参观校园。

—Her voice rings through the sky.

她的声音响彻云霄。

（2）加译：在英语介词短语前加汉语动词或形容词。

—the picture on the wall.

挂在墙上的画。

—the staircase to the lobby.

通向休息室的楼梯。

—In the West or throughout USA there is no place like Westlake.

在西欧或全美再也找不到西湖那样好的地方了。

—What is it beyond a joke?

这不是开玩笑，还能是什么呢？

（3）分译：将介词短语分译。

a．让步的分句

—With many such qualities, she remains modest and prudent.

尽管她有这么多长处，她仍然谦虚谨慎。

b．条件的分句

—On the gold standard, even fewer people than now would own most of the world's wealth, with the rest of mankind left bare-assed.

如果实行金本位，那么全世界的财富会集中到更少的人手中，而其余的人就会变成穷光蛋。

c．原因的分句

—The machine is working none the worse for its long service.

这台机器并不因为使用时间长而性能变差了。

d．目的的分句

—Now I want every house searched for firearms.

现在我要搜查每一所房子，把藏着的武器找出来。

（4）反译：某些介词反译比正译通顺。

—This machine is expensive beyond my reach.

这台机器很贵，我买不起（比“……，超出我力所能及的范围”好）。

—Until dawn, nothing could be done.

天不亮，什么也干不成。（比“直到天亮，……”好）

—We all went except him.

我们都去，就他不去。

—We all went besides him.

不只他去，我们也去。

（5）改译：把译意转换一下，使文句格外流畅。

—Education begins with a man's birth.

教育从一个人的出生就开始了。（不译成“教育随着……”）

—Shanghai is among the largest cities in the world.

上海属于世界上最大城市之列。（不译成“上海是世界上最大城市之中的一个”）

（6）不译：时间、地点或属性的介词可以不译。

—Smoking is prohibited in public places.

公共场所，禁止吸烟。

—The letter announced the death <u>of</u> his father.
这封信说他父亲死了。
—<u>In</u> the past, it was the custom <u>with</u> the peasants to marry earth.
过去，农民有早婚的习惯。
（7）特种改译：表示“程度”“次序”的介词可译成“比”。
—This is important <u>before</u> anything.
这<u>比</u>什么都重要。
—I bought the book <u>at</u> one half of the usual price.
我用<u>比</u>通常便宜一半的价钱买到了那本书。
—You are second <u>to</u> none.
你不<u>比</u>任何人差。

Part 8 How to Write Scientific Papers

本部分主要介绍英语科技论文的写作技巧，包括英语科技论文的主体结构、一般写作技巧、常用时态、人称和语态，以及常用表达方法等。

8.1 英语科技论文的主体结构及其要点

英语科技论文的整体组织结构和语言风格规范尤为重要，和建造楼房一样，写一篇论文也需要一份蓝图。英语科技论文蓝图最常见的是所谓的 IMRAD 结构，即：

- Introduction（引言）
- Methods（方法）
- Results and Discussions（结果与讨论）

采用 IMRAD 结构的论文具有简单、清晰、明了、逻辑性强的特点，因而这一结构被国内外学者广泛采用。采用 IMRAD 结构的论文，首先需要介绍研究课题的目的和意义，其次对研究工作所采用的研究方法、实验手段和材料进行描述，最后对研究结果及由此获得的主要结论进行总结和详细讨论。作者的任务就是根据 IMRAD 蓝图结构，将上述信息介绍给读者，以供读者参考。

论文标题（Article） 科技论文发表以后，最多被读者看到的是论文的标题。在期刊目录上，在互联网上，看到论文标题的读者可能会比最终看到全文的读者要多成千上万倍。因此，论文标题能否将大批潜在的读者吸引住，从而提高论文的影响力是至关重要的。撰写标题的要领在于尽可能用最少的词将论文最核心的内容表述出来。

作者和作者单位（Author Affiliation） 论文的署名表明作者享有著作权且文责自负，同时，作为文献资料，也便于日后他人索引和查阅。此外，论文署名还便于作者与同行或读者的研讨与联系，因此，有必要提供作者的身份、工作单位和通信地址，但标注时应准确、简洁。

摘要（Abstract） 摘要是论文的缩影，是对论文内容的简要概括和描述。摘要的作用是为读者提供关于文献内容的足够信息，包括：论文所涉及的主要概念和主要问题、采用的研究方法，以及得到的主要结论，以便于读者从摘要中了解作者所开展的主要研究活动。最重要的是，摘要可以帮助读者判断论文是否有助于自己的研究工作，是否有必要获取全文。

引言（Introduction） 引言的作用是提供足够的信息以明确研究工作背景和研究内容，通常应包括以下信息：（1）为什么写这篇论文，需要解决什么样的问题；（2）与研究课题相关的历史回顾和研究背景，进而明确本课题在学科领域中所占的地位及课题的研究意义和价值；（3）研究工作所涉及的界限、规模和范围；（4）研究工作的理论依据和采用的试验设备基础；（5）需要达到的预期目标；（6）研究涉及的相关概念和术语的定义。

上述内容不必逐一介绍，而要视具体情况进行取舍，以突出重点，但关键是要从一开始就吸引住读者的注意力，使读者了解选择这个课题的原因，以及这项研究的重要性等。

方法（Methods） 在撰写方法部分时，应注意要重点突出，详略得当。对公知公用的方法写明该方法名称即可；引用他人的方法、标准，或者虽有应用但尚未被人们熟知的新方法应注明文献出处，并对该方法进行简单介绍；对论文改进和创新的研究方法部分则应详细进行介绍。

总之，研究方法的介绍，既便于为其他研究者提供一个可重复研究的手段，又可以提高读者对论文研究工作及其结果可靠性的认可度。

结果和讨论（Results and Discussions） 结果是对研究中所发现的重要现象的归纳和总结，论文的讨论由此展开，论文对问题的判断和推理由此导出，全文的主要结论由此获得，因此，这一部分应该是论文的核心部分。

讨论是论文的重要部分，在全文中除摘要和结论部分外，讨论部分受关注度最高，是读者最感兴趣的部分，对读者很有启发作用，也是比较难写的部分。讨论部分的重点包括论文内容的可靠性、外延性、创新性和可用性。在讨论部分中，作者需要回答引言中所提出的问题，评估研究结果所蕴含的意义，并用结果去论证所提问题的答案。讨论部分撰写质量的好坏将直接影响作者对论文价值的判断。

结论（Conclusions） 科技论文的结论部分与引言部分是相呼应的。在结论部分中，应针对引言中提出的需要解决的问题及预期的目标做出明确的回答，是论文中继摘要和引言之后，从前瞻的角度，再次强调问题的重要性和研究的价值。

科技论文的结论部分紧跟在讨论部分的后面。读者往往在看过摘要之后，紧接着就会看结论部分，以了解研究工作的主要成果，再决定是否有必要认真地阅读全文或其中的一部分。由此可见，结论是论文研究成果的集中展现部分，作者应予以充分重视。

致谢（Acknowledgement） 在致谢部分应对除论文作者以外的那些对研究工作得以完成提供了帮助和有益讨论的人或部门进行感谢。此外，还需要对论文研究工作得到经费资助的基金项目进行鸣谢。

参考文献（References） 罗列出开展论文研究工作所参考过的相关文献，这些文献均需要在正文中进行标注，其罗列顺序应与正文中出现的顺序相一致。

8.2 英语科技论文的写作技巧

学术论文写作的主要目的是介绍科学知识、解释科技原理、分析解答自然界或人类社会的问题，它承担着传播知识、记载自然现象、传递科技发展信息及科学研究成果的重要任务。为了使其内容体现出科学性、准确性和客观性，科技论文的英语表达在措辞和句法结构方面有一些基本的规范与技巧，作者在撰写论文时应该了解和掌握。

1. 一般技巧

- 可适当强调研究中的创新、重要之处，以突出论文的创新特色，但尽量避免使用评价性的语言；
- 尽量使用简洁的句子，力求表达准确、简洁、清楚，以避免由于文字表述不清而引起读者对论文内容在理解上的困难；
- 充分展现论文的主要论点、主要结论和重要细节；
- 查询拟投稿期刊的作者须知，以了解其对论文形式的要求。

2. 时态

介绍背景资料时，如果句子的内容为不受时间影响的普遍事实，应使用现在时；如果句子的内容是对某种研究趋势的概述，则使用现在完成时。

- 在叙述研究目的或主要研究活动时，如果采用“论文导向”，多使用现在时（如：This paper presents …）；如果采用“研究导向”，则使用过去时（如：This study investigated …）。例如：

  —This article summarizes …

  —We investigated …

- 概述研究方法和主要结果时，通常用现在时。例如：

  —We describe …

  —Our results indicate that …

- 叙述结论或建议时，可使用现在时、臆测动词或 may、should、could 等助动词。例如：

  —We suggest …

3. 人称和语态

通常使用第三人称、过去时和被动语态；然而，为了简洁、清楚地表达研究成果，不应刻意回避第一人称和主动语态。

4. 常用表达方法

引言部分：

- 回顾研究背景　review, summarize, present, outline, describe
- 阐明写作或研究目的　purpose, attempt, aim
- 介绍论文重点内容或研究范围　study, present, include, focus, emphasize, attention 方法部分：
- 介绍研究或实验过程　test, study, investigate, examine, experiment, discuss, consider, analyze, analysis
- 说明研究或实验方法　measure, estimate, calculate
- 介绍应用、用途　use, apply, application

结果部分：

- 展示研究结果　show, result, present, imply, indicate
- 介绍结论　summary, introduce, conclude

讨论部分：

- 陈述论文的论点和作者的观点　suggest, report, present, explain, expect, describe
- 阐明论证　support, provide, indicate, identify, find, demonstrate, confirm, clarify
- 推荐和建议　suggest, suggestion, recommend, recommendation, propose, necessity, necessary, expect

总之，英语科技论文的写作要遵循多看、多写、多练的原则，在表述方面尽可能力求简洁，重点突出，关键是需要突出论文的创新特色，使读者能更好地了解论文在方法上的创新，以及取得的新结果和新结论，最大程度地为读者开展相关研究工作提供有用参考。

Grammar　专业英语翻译方法（八）：非限定动词的译法

1．不定式

主要作为名词，无人称、数、时态的变化，主要用作主、表、宾语，也可用作形容词和副词。

—To serve the people is glorious.

为人们服务是光荣的。（主语）

—She likes to read poems loudly.

她喜欢朗诵诗歌。（宾语）

—The important things is to be good at learning.

重要的问题在于善于学习。（表语）

—I come to attend the conference.

我是来参加会议的。（副词做状语）

2．动名词

主要作为名词，无人称、数的变化，有时态、语态变化。主要用作主、表、宾语，也可用作定语。

—Saving is getting.

节约就是增收。（主语和表语）

—Would you mind showing me the way?

麻烦你告诉我怎么走，好吗？（动词宾语）

—They insist on my staying a week longer.

他们坚持要我多待一星期。（介词宾语）

—Take some sleeping pills, you'll soon fall asleep.

服几粒安眠药，你很快就会睡着。（定语）

3．分词

主要作为形容词，无人称、数的变化，有时态、语态变化，分为现在分词和过去分词。主要用作定语、表语，也可用作状语。

—The working class is the leading class.

工人阶级是领导阶级。（定语）

—I saw him taking the book away.

我见他带走了那本书。（宾语补语）

—The football match is quite exciting.

这场足球赛相当精彩。（表语）

—Turning to the right, you will find a path leading up to the park.

向右转弯，你可顺着一条小路去公园。（状语）

4．现在分词与过去分词的比较

其区别在于：

时态：现在分词——正在进行的动作
过去分词——已经完成的动作
语态：现在分词——主动语态
过去分词——被动语态
boiling water　正在开的水
boiled water　开过的水
developing countries　发展中国家
developed countries　发达国家
现在分词用作表语表示主语的性质，过去分词用作表语表示主语的状态。
—The story that she tells us is extremely touching.
她所讲的故事非常动人。
—He looks somewhat confused.
看来他有点茫然。

5．动名词与现在分词用作定语时的区别

前者表示目的，后者表示性质。
—a sleeping car　卧（铺）车（厢）（动名词）
—a sleeping child　睡着的小孩（现在分词）
—drinking water　饮用水（动名词）
—a drinking horse　饮水的马（现在分词）

6．动名词与现在分词作表语时的区别

前者表示动作的结果，后者表示事物的性质或状态。
—At present, the most urgent measure is saving power and energy.
目前急需采取节约能源的措施。（动名词）
—His report is quite convincing.
他的报告很有说服力。（现在分词）

7．动名词与不定式用法的区别

（1）动名词表示一般、抽象行为，不定式表示特殊、具体行为。
—I like talking with workers.
我喜欢和工人谈话。
—I like to talk with the workers about the production methods.
我喜欢和工人谈生产方法问题。
（2）动名词表示旁人的动作行为，不定式表示自己的行为动作。
—I like singing.
我喜欢听唱歌。
—I like to sing.
我想唱歌。
（3）动名词和不定式在某些动词（stop, forget, remember）等的及物和不及物情况下有不同的含义。

—He stopped talking.

他停止谈话。

—He stopped to talk with me.

他停下来和我谈话。

（4）动名词在时间上表示已经进行或正在进行的动作，不定式表示准备进行或即将进行的动作。

—I forgot writing the letter.

我忘记了写过那封信。

—I forgot to write the letter.

我忘了写那封信。

8．动名词与不定式做复合宾语时意义不同，动名词强调动作的延续性，不定式则表示一般动作

—I saw the workers walking along the beach.

我看见工人们正在沿着海滩散步。

—I saw the workers walk along the beach.

我看见工人们沿着海滩散步。

References

[1] M.Young. Optics and Lasers: Including Fibers and Optical Waveguides (5th Edition). Springer-Verlag Berlin Heidelberg, 2000.

[2] Leo Levi. Applied Optics: A Guide to Optical Design / Volumn 2. Published by Wiley, Canada, 1980.

[3] W.Koechner. Solid-State Laser Engineering (Six Revised and Updated Edition). Springer, U.S.A, 2006.

[4] Govind Agraawal. Nonlinear Fiber Optics, Third Edition(Optics and Photonics). Academic Press, 2001.

[5] Amnon Yariv. Optical Electronics in Modern Communications (5th Edition). Publishing House of Electronics Industry, 2004.

[6] Max Born, Emil Wolf. Principles of Optics (7th Edition). Cambridge University Press, 1980.

[7] P.Hariharan. Optical Holography: Principles, techniques and applications (2th Edition). Published by the Press Syndicate of University of Cambridge, 1996.

[8] Ting-chung Poon. Digital Holography and three-Dimensional Display: Principles and Applications. Springer, U.S.A, 2006.

[9] Peter W.Milonni, Josph H.Eberly. Laser Physics. New Jersey, 2010.

[10] Rajiv Ramaswami, Kumar N.Sivarajan, Galen H.Sasaki. Optical Networks: A Practical Perspective (3th Edition). Morgan Kaufmann Publishers, U.S.A, 2010.

[11] H Angus Macleod. Thin-Film Optical Filters (3rd Edition). Institude of Physics Publishing Bristol and Philadelphia, 2001.

[12] Levinson, Harry J. Principles of Lithography (3rd Edition). Published by SPIE, U.S.A, 2010.

[13] Atanu Biswas, Sujay Datta et al. Statistical Advances in the Biomedical Sciences: Clinical Trials, Epidemiology, Survival Analysis, and Bioinformatics. Published by John Wiley&Sons, Inc., Hoboken, New Jersey, 2008.

[14] Enrico Forestieri. Optical Communication Theory and Techniques. Springer, Boston, 2005.

[15] Joseph W.Goodman. Introduction to Fourier Optics (2nd Edition), 1996.

[16] Jan P.Allebach, John D.Baloga *et al.*.Handbook of Optics/Volume III, Classical Optics, Vision Optics, X-Ray Optics(3rd Edition). McGraw-Hill Proffessional, 2009.

[17] Eugene Hecht, Optics(4th Edition). Adelphi Unversity, 2001.

[18] 钟似璇. 英语科技论文写作与发表. 天津：天津大学出版社, 2004.

[19] 辜嘉铭. 英语科技论文写作精要. 武汉：武汉大学出版社, 2006.

[20] 李霞, 王娟. 电子与通信专业英语. 北京：电子工业出版社, 2010.

[21] 金志权, 张幸儿. 计算机专业英语教程. 北京：电子工业出版社, 2010.

[22] Huard S. Polarization of Light. Wiley-VCH, 2002.

[23] Shulika, Oleksiy, Sukhoivanov, Igor. Advanced Lasers: Laser Physics and Technology for Applied and Fundamental Science. Springer, Netherlands, 2015.

[24] Yun-Shik Lee. Principles of Terahertz Science and Technology. Springer, New York, NY, 2008.

[25] Daryoosh Saeedkia, Handbook of terahertz technology for imaging, sensing and communications. Woodhead Publishing, Canada,2013.

[26] T. Y. Chang, Kenneth J. Button. Reviews of Infrared and Millimeter Waves: Volume 2 Optically Pumped Far-Infared Laser. Springer, US, 1984.

[27] Saroj Rout, Sameer Sonkusale. Active Metamaterials: Terahertz Modulators and Detectors. Springer International Publishing, 2017.

反侵权盗版声明

举报电话：（010）88254396；（010）88258888
传　　真：（010）88254397
E-mail：dbqq@phei.com.cn
通信地址：北京市海淀区万寿路 173 信箱
　　　　　电子工业出版社总编办公室
邮　　编：100036